Practica

Readers in Experimental Philosophy at the
BOSTON ATHENÆUM
(1827–1850)

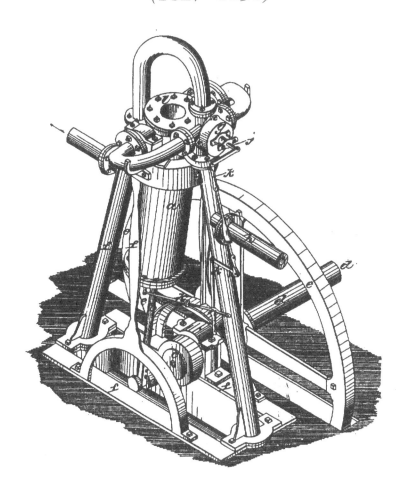

SCOTT B. GUTHERY

DOCENT PRESS
Boston, Massachusetts, USA
www.docentpress.com

Docent Press publishes books in the history of mathematics and computing about interesting people and intriguing ideas. The histories are told at many levels of detail and depth that can be explored at leisure by the general reader.

Cover design by Brenda Riddell, Graphic Details. The figure is taken from US 4749½X, the patent granted to Boston ironmonger and Boston Athenæum member Ebenezer Avery Lester on May 14, 1827 for his Pendulum Steam Engine.

Produced with TEX. Typeset in Minion.

9 8 7 6 5 4

ISBN-10: 1-942795-94-7
ISBN-13: 978-1-942795-94-0

To Mary, my partner in all dances, plain and fancy.

Contents

List of Figures

List of Tables

List of Equations

Foreword

How should a library measure its value to society? By the size and usefulness of its collection? By the community it fosters? The thinkers it connects? The new knowledge it helps to generate? By all these measures, the Boston Athenæum fulfilled its mission during the years considered in this volume: 1827 to 1850.

The author brings a patient, inquisitive, informed eye to his study of titles in science and technology borrowed by Athenæum readers during this expansive period. Proceeding with a mathematician's precision, Guthery lays out his methodology, plots the data, spots patterns, and interprets those patterns with gusto. The resulting text defies conventional genres, combining a historian's familiarity with antebellum New England, with a mathematician's ability to test the specific equations used by his historical subjects, with a gifted teacher's deployment of lucid analogies to elucidate complex ideas.

The approach seems just what the subject demands. The style, brimming with the thrill of discovery, and enlivened with witty observations, keeps the reader turning pages, discovering networks of practical researchers and entrepreneurs who borrowed books (for themselves and their family members) related to specific undertakings—before, in some instances, those readers-inventors-investors wrote new titles based on their findings, works that would come to stand alongside the others on the Athenæum's shelves.

But *Practical Purposes* is more than an accumulation of intriguing revelations about specific patent applications, engineering experiments, medical innovations, and pedagogical methods. Considered as a whole, this book challenges our assumptions about a particular time and place in the history of science. Guthery questions the conventional wisdom that New England did not generate scientific knowledge in this era, and determines that while Boston was not a center for what we would consider to be academic science, it was a hotbed of what was then called experimental philosophy—the practical application of scientific and technological understanding. In that sense, the Boston Athenæum could be regarded as having performed, in the second quarter of the nineteenth century, aspects of the role of a contemporary technology incubator.

The circumstances, as Guthery demonstrates, were just as the founders had envisioned:

> "The Athenaeum may be recommended as a place of social intercourse. But it will principally be useful as a source of information, and a means of intellectual improvement and pleasure. It is to be a fountain, at which all, who choose, may gratify their thirst for knowledge." (Dr. John Thornton Kirkland, "Memoir of the Boston Athenaeum, with the Act of Incorporation, and Organization of the Institution," 1807)

Elizabeth E. Barker
Stanford Calderwood Director
February 15, 2017

Acknowledgments

Research that casts its net widely, such as that reported here, necessarily calls on expertise far beyond the investigator's own. The assistance of some of these experts is acknowledged below. These were by no means the only individuals who were gracious with their time and knowledge. I acknowledge the help of others and apologize for not doing so by name.

At the top of the list of unnamed contributors are the men and women who worked at the Boston Athenæum's circulation desk from 1827 to 1850. These are the people who day-after-day carefully recorded the borrowing and returning of books and journals in the Athenæum's books borrowed registers, creating the foundation for this work and future research.

Which isn't to say that the book hasn't benefited in many ways from the guidance and generosity of everyone at the Boston Athenæum today. Of particular note is the patience and persistence of Patricia Boulos, the head of the Athenæum's digital programs. In addition to digitizing the Athenæum's books borrowed registers in the first instance, Patricia cheerfully fielded requests for high-resolution images of selected pages in the registers as well as pages in books in the Athenæum's collection.

Carolle Morini, the Boston Athenæum's archivist, curated the registers and helped me find and parse the minutes of the various Athenæum library committees. Mary Warnement, Head of Reader Services, Catharina Slautterback, Curator of Prints and Photographs, and Dani Crickman, Children's and Young Adult Services Librarian, were forever patient with my questions about the history of the Boston Athenæum as well as the history of early American libraries and literature.

Scott Sagar, Curator of Collections at the Lycoming County Historical Society, as well as Tom Drogalis and John Bambrick of the Schuylkill County Historical Society, helped me find my way around the Eastern Pennsylvania coal fields. Peter H. von Bitter, Senior Curator Emeritus, Natural History, Royal Ontario Museum, and Professor Emeritus, Earth Sciences, University of Toronto, generously shared his research files regarding the tussle between the Canadian geologist, Abraham Gesner, and Athenæum readers Francis Alger and Charles T. Jackson.

Amy Ackerberg-Hastings plumbed the Uriah Boyden archives at the National Museum of American History with a sharp eye for Boyden's mathematical skills. Marcis Kempe, retired director of the Massachusetts Water Resources Authority operations support unit and unofficial historian of Boston's water system, helped me look behind the reports to Boston's many water committees. Karen Graham, a specialist at the Massachusetts Water Resources Authority library, searched the MWRA's archive for early maps of Boston's water system. Beth Carroll-Horrocks, Head of Special Collections for the Commonwealth of Massachusetts, never failed to find historic needles in the haystack of government documents. Janet Steins, Associate Librarian for Collections and Research Librarian at Harvard's Tozzer Library, helped out with statistics of Harvard's book collections.

Long-time friend and colleague, Tim Jurgensen, explained in words I could understand the intricacies of calorimetric measurement. Fellow member of the Athenæum's Mathematics, Technology, and Society discussion group, Tom Pascarella, gently set me straight on some points in the history of chemistry.

The contributions of readers of early drafts are less apparent but equally important in crafting the words on the page. Should the reader take particular delight in a paragraph or passage it is highly likely that Marie Oedel, Elizabeth Barker, Jordan Goffin, or Gerard Kiernan helped make it so.

In spite of the expertise and unselfish kindness of the above individuals, there will inevitably be slips. I take full responsibility for these.

Preface

The Influence of the Boston Athenæum

Over a century ago Barrett Wendell paged through the Boston Athenæum's books borrowed registers while tallying the contributions readers at the Athenæum had made to the humanities. In the resulting essay, "The Influence of the Athenæum on Literature in America," Wendell observed:

> For a great many years the simple custom of the library was to write down in a book just who took volumes away to read, and how long he kept them. No trace remains of what pages these men and women turned in the pleasant seclusion of the library; but by inspecting these brief records of what they carried home, we can form some notion of what they sought and found in the cordial surroundings where the friendly inspiration of letters so surely lingers. [20, pp. 6–7]

My objective is the same as Wendell's—to inquire as to what the books Athenæum members borrowed can tell us about the influence they had on their community—but with three caveats. First, I am concerned with readers of the scientific and technical literature. Second, I am interested in connections between what these individuals read and what they built, as well as what they wrote. And finally, I consider what their preference among books on a particular topic—which I take to be a preference for one author's style of discourse over another's—tell us about what these readers sought and valued in the scientific and technical literature.

Some of the readers whose names appear in the registers and on the following pages influenced science and engineering in the same way that readers in the humanities influenced the arts: they wrote books and journal articles. Josiah Cooke wrote about chemistry. Jacob Bigelow wrote about botany. And Edward Tuckerman wrote about lichens.

The influence of other readers is, however, to be found in what they built rather than in what they wrote. Uriah Boyden built turbines. Erastus Bigelow built looms.

And Alexander Parris built lighthouses. To calibrate the influence of these readers one must look into the manner of thinking that went into what they built. It is to this end that I examine their preferences in expository style in scientific and technical discourse.

Readers in Experimental Philosophy

Experimental philosophy is the science of everyday life, distinctly and deliberately as lived outside of the academy. It is science and technology at a human scale as observed and understood by the man in the street. In the words of a newspaper article that appeared on page 2 of the September 7, 1822, edition of the *The Independent Bostonian*:

> Experimental philosophy has its foundation in experience wherein nothing is affirmed or assumed as a truth but what is founded upon occular [*sic*] demonstration, or which cannot be denied without violating common sense or perception.

As used in this book, experimental philosophy is more specifically the quantitative discipline that discovers and characterizes regularities in nature on which one can rely and build. In short, the phrase 'reader in experimental philosophy' is taken to mean a reader in science and technology for practical purposes.

There is an extensive body of research using records of books borrowed[1] from private and public libraries. Many of these studies sought to gain insight into historical contexts and social norms.[2] They moved from a characterization of a corpus of books borrowed—by a person, a family, a community, etc.—to hypotheses about the mores, attitudes, opinions, and interests of the borrowing population. The methodology of these studies was to use books borrowed data to draw conclusions about what people thought and believed as reflected in their preferences in reading material.

Any use of readership data is haunted by a simple question: "Did anyone actually read and understand the book?" Wendell acknowledged the lurking presence of this question: "The fact that people signed their names for books need not mean that they ever actually read them." [20, p. 11] Owen Gingerich's *The Book Nobody Read: Chasing the Revolutions of Nicolaus Copernicus* [111] is the tale, both fascinating and cautionary, of his decade-long quest to answer this question for just one book.[3] Gingerich used marginalia to speak to the question. I will use what people didn't read. Let me explain.

---

[1] ...also known as charging records, circulation records, loan records, library borrowers' registers, due date cards, etc.

[2] Examples are [189], [108], and [101]. With particular regard to colonial and antebellum New England, see the work of Ronald and Mary Zboray including [262], [263], [264], [265], [266], and [267].

[3] *De Revolutionibus Orbium Coelestium* by Nicolaus Copernicus.

While the Boston Athenæum was one of the five largest libraries in the United States during the period of this study, the number of books on the shelf on any particular scientific or technical topic was small enough that one can ask with some hope of an answer "Why did readers prefer this book over the others on the same topic and what might this preference tell us about the reasons behind their reading?" It is argued in the following that readers in experimental philosophy at the Boston Athenæum during the period of the study read primarily for practical purposes and, as a consequence, they preferred books and journals that were directly and reliably applicable to a purpose at hand. It is further proposed that the expository style of these preferred texts was the experimental essay.

An experimental essay is a way of writing championed by the Irish chemist, physicist, and inventor Robert Boyle. Boyle used it himself to report on his own experiments and he advocated its use by fellow members of the Royal Society. He described the experimental essay in a number of his works, for example:

> [W]hereas by the way of writing, to which I have condemned myself, I can hope for little better among the more daring and less considerate sort of men, should you shew them these papers, than to pass for a drudge of greater industry than reason, and fit for little more, than to collect experiments for more rational and Philosophical heads to explicate and make use of. But I am content, provided experimental learning be really promoted, to contribute even in the least plausible way to the advancement of it; and had rather not only be an under-builder, but even dig in the quarries for materials towards so useful a structure, as a solid body of natural philosophy, than not do something towards the erection of it. [40, p. 307]

The epigram that appears at the beginning of each chapter is taken from Boyle's *Proëmical Essay*, his treatise devoted to the nature and advantages of the experimental essay.

Notes on Methodology

In determining if an entry in a books borrowed register was included in the study and, if so, to assign it to one of the subject matter categories under consideration, the following general rules were followed:

- A reasonable association of an entry in the register with an entry in one of the Athenæum's catalogs had to be established. It is granted that 'reasonable' is a subjective construct. A registry entry that consisted of nothing more than 'Biot', for example, was highly likely to be to a book written by Jean-Baptiste Biot, but as there are numerous scientific and engineering texts by Biot in the catalogs, no association with a particular title could be established so this registry entry would not be included in the study. Had there been only one book by Biot in the catalogs, however, it would have been included.

- Save for two journals and *Elements of Technology* by Jacob Bigelow, an entry is included in the study only if it could with high certainty be placed in one of the fourteen subject matter categories. For example, the register entry "Bakewell's Geology" would be taken to be *Introduction to Geology* by Robert Bakewell assigned to the geology subject matter category .

- Entries referencing a sub-domain of a subject matter category are included in the more general category. For example, the entry "Brooke's Crystallography" was placed in the mineralogy category.[4]

- Entries for titles in the philosophy of science, such as Whewell's *Inductive Science*, and compendiums of scientific topics, such as Rees' *Cyclopædia*, are not included in the study on the grounds that it would be difficult to associate reading in these titles with an interest in a particular scientific or technical subject.

- Books in fields which entailed little experimental activity such as geography and gardening or which showed very few register entries such as archaeology, climatology, animal magnetism, and phrenology are not included. A small number of Athenæum readers did engage in experimental agriculture as documented by Tamara Thornton in *Cultivating Gentlemen* [229]. The intersection of agriculture and entomology is covered in Chapter 9.

- The Boston Athenæum absorbed the Boston Medical Library and its membership early in the period covered by the study and as a result there are a large number of charges of medical books in the registers. While these titles are likely to have been read for practical purposes and to contain experimental content, they have been excluded from the study due to my own lack of background in the history of medicine. These register entries warrant a study of their own.

An attempt was made to detect and not count renewals. If a register entry consisted solely of ditto marks and the charge date on the entry was close to the due date of the referenced entry, then the charge was regarded as a renewal and not counted. Errors in the application of this rule are certain and, for the most part, are in the direction of not counting charges that should have been counted. This is particularly true for periodicals, so their charge counts are almost certainly too low.

---

[4]It is acknowledged that making mineralogy its own category is a twenty-first century point-of-view since at the time mineralogy was generally regarded as a sub-domain of chemistry.

The exact charge date was recorded for approximately 38% of the 3,761 register entries in the study. These are summarized by subject matter category in the table below. The charge date for the remaining entries was taken to be halfway between the charge date of the first and last entries on the register page.

Category	Charges
Astronomy (*Cosmos*)	72
Chemistry	459
Geology	445
Pyronomics	130
Hydraulics	161
Mechanics (Railroads)	63
Mineralogy	123
Total	1,453

Categories with Exact Charge Date

There are undoubtedly titles in the Athenæum's books borrowed registers that some would argue should be included but aren't. As well, there are titles that have been included that others would question. I believe the conclusions drawn would be unaffected by addressing these shortcomings but that, of course, is a determination left to the current reader.

Scott B. Guthery
Boston, Massachusetts
February, 2017

And you will easily pardon me the injury, which for your sake I do my own reputation by this naked way of writing, if you, as well as I, think those the profitablest writers, or, at least, the kindest to their perusers, who take not so much care to appear knowing men themselves, as to make their readers such.

A Proëmical Essay
Robert Boyle

Chapter 1

Reading for Practical Purposes

1.1 Kirk Boott and a Meeting of the Directors of Merrimack Manufacturing

On Monday, January 18, 1830, Patrick Tracy Jackson asked Kirk Boott to "to call a meeting of the Directors [of the Merrimack Manufacturing Company]…at which I shall propose…a meeting of the Proprietors of that stock to the project for building a railroad from [Boston] to Lowell."[1] The meeting was hastily scheduled for the end of the week. On the day of the meeting, Friday, January 22, Boott borrowed four books from the Boston Athenæum on the planning, building, and operation of a railroad. Figure 1.1 is a clip from Volume I of the Boston Athenæum's books borrowed register showing the charges to Boott's account.

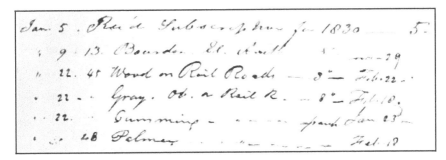

Figure 1.1: Books Borrowed by Kirk Boott on January 22, 1830[2]

---

[1] As quoted in [41, p. 4].
[2] Readers with an interest in mathematics might note that Kirk Boott borrowed Bourdon's *Elémens d'Arithmétique* on January 9.

The four books Boott borrowed to prepare for the meeting were the following:

- Nicholas Wood, *A Practical Treatise on Rail-Roads* [252]
- Thomas Gray, *Observations on a General Iron Rail-way* [120]
- T. G. Cumming, *Illustrations of the Origin and Progress of Rail and Tram Roads, and Steam Carriages, or Loco-motive Engines* [68]
- Henry R. Palmer, *Description of a Railway on a New Principle* [185]

A railroad between Boston and Lowell would compete directly with the Middlesex Canal which had opened in 1802. The canal was operated by the Middlesex Canal Corporation, and was, by all accounts, a commercial success so Boott and his associates could expect some vigorous push-back from the canal's investors. While profiting from moving goods between Lowell and Boston was what was really at stake, arguments in favor of a railroad would have to rest on the translation of the technical and commercial aspects of the railroad into investor returns and public benefits. Boott's task was to convince the directors and eventually the Massachusetts legislature that the Boston and Lowell railroad should be built.

In the introduction to his *A Practical Treatise on Rail-Roads*, Wood[3] says precisely how he will approach his subject:

> The greatest care has been used in the prosecution of the different experiments, and the most minute details are given, in order that the reader may be able to judge of the credit to which they are entitled; our object has been to furnish practical data on the subject, and in doing so, not to assume any theory, or deduce any proposition, which is not supported by experiment. [252, p. vi]

Wood is going to lay the outcomes of experiments he conducted before the reader to support his discourse. He is going to describe these experiments in sufficient detail that the reader can repeat them. His discourse will neither assume or propose any theory. And finally, his report will be written in a way that the outcomes can be put to work for practical purposes. In short, the data will speak for itself. If you don't believe it, you can perform the experiment yourself.

The directors of Merrimack Manufacturing including Boott and Jackson were not babes in the woods when it came to industrial-scale, technology-driven businesses.

---

[3] As testimony to his expertise in railroading technology, Nicholas Wood was one of three judges at the famous Rainhill Trials sponsored by the Liverpool & Manchester Railway in October, 1829, to pick a steam engine to power the railroad. Dame Hubris was also in attendance. Shortly before the trials, Wood expounded "It is far from my wish to promulgate to the world that the ridiculous expectations, or rather professions of the enthusiastic speculatist [sic] will be realized, and that we shall see engines travelling at the rate of twelve, sixteen, eighteen, or twenty miles per hour." As it turned out, all three entrants in the trial—Rocket, Novelty, and Sans Pareil—topped twenty miles per hour at one point or another and Rocket topped thirty during its victory lap.

Starting with $600,000 and a swamp-choked canal on the Merrimack River they had built and were operating the massive Lowell textile mill complex. They had begun with a prototype mill in Waltham to familiarize themselves with water wheel power and spinning technologies as well as the engineering problems they might face when they amped up the scale of the technology to a mill complex. They knew business start-ups and they knew engineering project management. This time, however, there would be no practice railroad. The directors of the Merrimack Manufacturing Company were going to place a serious-money bet on the experimental outcomes reported on in the books borrowed by Boott.

1.2   Experimental Philosophy in Antebellum New England

While an anachronism today, the term 'experimental philosophy' was frequently used without explanatory text in antebellum New England.

In medicine, for example. A qualification which had to be met to be granted a licence to practice medicine in Massachusetts stipulated in "An Act regulating the practice of Physic and Surgery" (passed 1819) was that the candidate...

> ...shall have an acquaintance with the latin language as is necessary for medical or Surgical education, and with the principles of geometry and experimental philosophy.[4]

And in the schools. In the preface to *The Boston School Compendium of Natural and Experimental Philosophy*, the author, Richard Green Parker, lists the advantages of his method of teaching science, one of which is that...

> ...the pupil can never be at a loss to distinguish the parts of a lesson which are of primary importance; nor will he be in danger of mistaking theory and conjecture for fact. [187, p. 6]

As well, Sir Richard Phillips writes[5] in the preface to his *Easy Grammar of Natural and Experimental Philosophy for the Use of Schools*:[6]

> Every instructor of youth must be aware that mere disquisition is of no use in the art of teaching, and that no science can be taught, if the student does not *work* or perform operations in it. [16, p. iv]

---

[4] As reported in the *Hampshire Gazette* of October 26, 1819, as well as other newspapers of the day.
[5] ...as David Blair.
[6] This 32mo pocketbook covers physics, mechanics, hydraulics, optics, astronomy, electricity, and magnetism in just 119 pages. Every chapter includes numerous learning experiments. The back matter consists of thirty-one pages of review questions and ten-page glossary.

And on the lecture circuit. The advertisement in Figure 1.2 appeared in the Thursday, May 30, 1843, edition of the *Boston Evening-Post*.

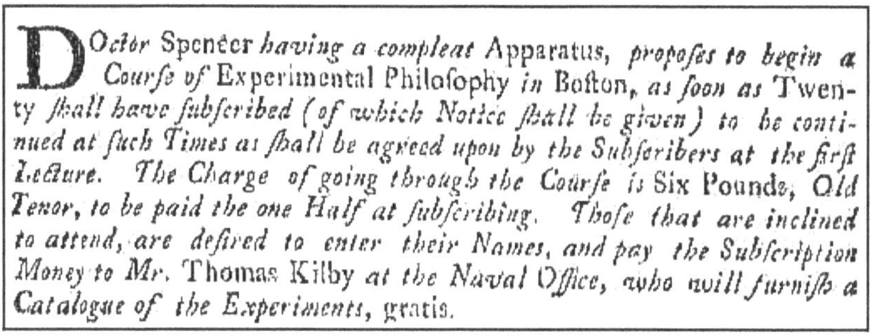

> **D**octor Spencer *having a compleat* Apparatus, *propofes to begin a Courfe of* Experimental Philofophy *in* Bofton, *as foon as* Twenty *fhall have fubfcribed (of which Notice fhall be given) to be continued at fuch Times as fhall be agreed upon by the Subfcribers at the firft Lecture. The Charge of going through the Courfe is* Six Pounds, *Old Tenor, to be paid the one Half at fubfcribing. Thofe that are inclined to attend, are defired to enter their Names, and pay the Subfcription Money to Mr.* Thomas Kilby *at the Naval Office, who will furnifh a Catalogue of the Experiments, gratis.*

Figure 1.2: Experimental Philosophy in an 1843 Course Advertisement

And finally in classified advertisements. The solicitation in Figure 1.3 appeared in the Sunday, January 20, 1743, edition of the *Boston News-Letter*.

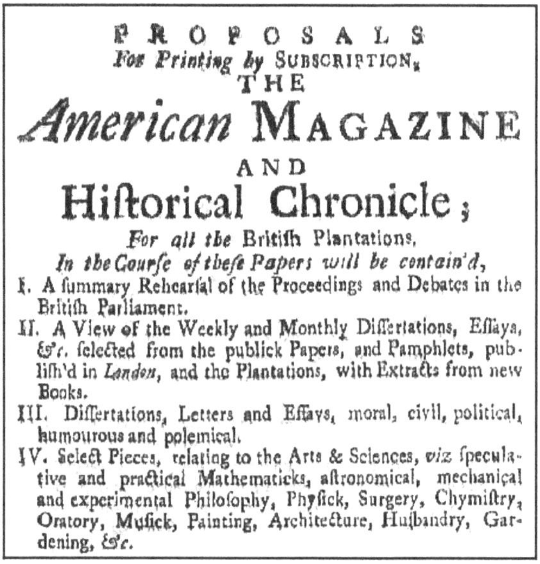

> **P R O P O S A L S**
> For Printing *by* SUBSCRIPTION,
> T H E
> *American* MAGAZINE
> A N D
> Hiftorical Chronicle;
> For all the Britifh Plantations.
> *In the Courfe of thefe Papers will be contain'd,*
> I. A fummary Rehearfal of the Proceedings and Debates in the Britifh Parliament.
> II. A View of the Weekly and Monthly Differtations, Effays, &c. felected from the publick Papers, and Pamphlets, publifh'd in *London*, and the Plantations, with Extracts from new Books.
> III. Differtations, Letters and Effays, moral, civil, political, humourous and polemical.
> IV. Select Pieces, relating to the Arts & Sciences, viz fpeculative and practical Mathematicks, aftronomical, mechanical and experimental Philofophy, Phyfick, Surgery, Chymiftry, Oratory, Mufick, Painting, Architecture, Hufbandry, Gardening, &c.

Figure 1.3: Experimental Philosophy in a 1743 Magazine Solicitation

As for books, Table 1.1 lists some titles in experimental philosophy in the Boston Athenæum's collection which were published before the period of this study.

Author	Short Title	Year
Desaguliers	*Lectures of Experimental Philosophy*	1719
Desaguliers	*Course of Experimental Philosophy*	1763
Green	*Course of Lectures on Experimental Philosophy*	1782
Vancouver	*General Compendium of Chemical, Experimental, and Natural Philosophy*	1785
Adams	*Lectures on Natural and Experimental Philosophy*	1794
Enfield	*Institutes of Natural Philosophy, Theoretical and Experimental*	1799
Young	*Syllabus of a Course of Lectures on Natural and Experimental Philosophy*	1802
Cavallo	*Elements of Natural or Experimental Philosophy*	1803
Blair	*Easy Grammar of Natural and Experimental Philosophy*	1814
Millington	*Epitome of the Elementary Principles of Natural and Experimental Philosophy*	1823

Table 1.1: Titles in Experimental Philosophy in the Athenæum's Collection

Experimental philosophy as tacitly understood in all these New England contexts and as found in the following chapters is not applied science as this term is understood today, namely the application of scientific theories to practical problems. Nor does it refer to the experimentation done in laboratories to divine those theories. It is a different sort of science. It is a science informing and being informed by the pursuit of practical purposes. It could be called street-smart science. When the term 'experimental philosophy' is used herein it is used in this sense.[7]

1.3 Purposeful Reading

By and large people buy and borrow books whose content interests them. There are exceptions, of course: buying a book as a gift, borrowing a book for a friend, collecting

---

[7]Wikipedia records that "Experimental philosophy is an emerging field of philosophical inquiry that makes use of empirical data in order to inform research on philosophical questions." 'Experimental philosophy' as used herein isn't this experimental philosophy either.

a book for its binding, and acquiring a book as an investment to name just a few. But these exceptions are statistical noise when considering the vast bulk of books bought and borrowed. This rather obvious property of the book trade, which has been in force at least since Gutenberg decided to print the Bible rather than *Tales of the Black Forest*, has been fruitfully harnessed in numerous analyses of private book collections, library catalogs, and Amazon rankings.

Knowing about a book can tell us about its readers. But it works the other way around too. Knowing about the readers can tell us about the book. That a book read by all the students in a chemistry class was a chemistry book would not be a great revelation. But if there are two books about chemistry on the shelf and one is consistently chosen in lieu of the other, then we are led to ask what reader needs and expectations underlay this discrimination. Uncovering these might expose a difference between the two books that was not apparent by just looking at the two books side-by-side.

Boston was expanding rapidly in the second quarter of the nineteenth century. The literary and industrial center of gravity of the new nation was drifting northward from Philadelphia. Boston's population was growing apace, its docks handled a brisk import/export trade, and charters for new businesses occupied no small part of the legislature's calendar. The members of the Boston Athenæum were thoroughly vested in all aspects of the city's cultural and commercial development. Keeping abreast of the latest developments in science and technology was not simply a matter of being culturally well-rounded; it was a matter of informed community involvement and a competitive edge in commercial ventures.

In the following chapters, I analyze science and technology titles recorded in the first four volumes of the Boston Athenæum's books borrowed registers in order to explore what the books that were borrowed say about their readers and what the readers' choices say about the books they borrowed. The primary method used to conduct this research is to strive to connect what people were reading with what they were doing.

An obvious assumption underpinning such an exploration is that titles in science and technology were being read for a purpose rather than for amusement. A second assumption at work is that the characteristics of a book's content for the purpose of understanding the book's readers—the organization of the subject matter, the manner of exposition, the nature of the illustrations and examples, and so forth—are conscious decisions made by the book's author. Thus, when we ask what a book says about its readers, we are implicitly asking how has the author of the book characterized and then written to the needs and expectations of the book's intended audience. Thus, a book is thought of as consciously and explicitly reflecting the author's view of how his intended audience wants to engage his subject.

The connections between books and readers at the Boston Athenæum was particularly strong in the institution's early years for two reasons. First, a noticeable number of the books in the Athenæum's growing collection were donated by members from their private collections. While it is common today to drop off paperbacks at the local public library that are no longer of any interest, books donated to the Boston Athenæum were typically of considerable value and not infrequently of continuing interest to the donor. A book donation was offered and accepted explicitly because the book was thought to be of timely interest to other members.

Second, the commercial and social ties between readers at the Athenæum were tighter and more numerous than such ties were between readers at, say, the Harvard library. This is the case because the Athenæum was, by design and intent, a social milieu not just collection of stacks. Books were purchased by the Athenæum's librarians and the Athenæum's book committee to reflect recognized and active communities of interest. By and large members of the Boston Athenæum were well enough off to purchase books that suited their strictly personal interests themselves. The titles in the books borrowed registers reflect shared interests.

Let's return to the four books Kirk Boott borrowed to prepare for the director's meeting.

Cumming's book on the progress of railroads is mostly about the competition between railroads and canals in Britain but it does review experiments with horse-drawn wagons as a preface to discussing "steam carriages or loco-motive engines." The cost of horses, particularly the cost of their care and feeding, was a common starting point for an economic comparison of canals and railroads. If one horse can pull more tonnage per day on a railroad than on a canal then score one for the railroad. For example, Cumming reports that...

> ...the most comprehensive and exact experiment that we have heard of, is stated to have been made by Joseph Wilkes, Esq. of Measham, in Derbyshire: the result is, that one horse, value 20*l.* on a rail-way declining at the rate of one foot perpendicular to one hundred and fifteen, the length of the road, drew twenty-one carriages, or waggons, laden with coals and timber, amounting in the whole to thirty-five tons, overcoming the vis inertiæ repeatedly with great ease. [68, p. 60]

Wilkes summarized his experiments on horse-powered railways with a simple formula. To move a load of X pounds up a grade of 1-in-K, Wilkes says, requires

$$\frac{X}{94} + \frac{X}{K} \tag{1.1}$$

pounds of traction.

The first term of Wilkes' formula, $X/94$, is the power needed to overcome friction and inertia and get the wagon rolling. The second term, X/K, is the power needed to move the load up the grade. The more gentle the slope, the bigger the value of K, the smaller the value of X/K, the less effort needed to move the wagon.

Gray's book is in the same vein as Cumming's,[8] singing the praises of steam-powered railroads. After noting that horses wear out faster than steam engines, Gray makes the following rather creative argument in favor of steam power:

> It has been calculated that there are at least 10,000 of these machines [steam engines] at this time at work in Great Britain; performing a labour more than equal to that of 200,000 horses, which, if fed in the ordinary way, would require above 1,000,000 acres of land for subsistence; and this is capable of supplying the necessaries of life to more than 1,500,000 human beings. [120, p. 89]

Gray uses the tabular display shown in Table 1.2 to summarize his results but his table serves the same purpose as Wilkes' formula. Gray's tabular display compares "the resistance of a rail-way with that of a canal or arm of the sea, in a calm atmosphere."

MPH	Ship	Railway
4	133	102
6	300	105
8	533	109
12	1200	120
16	2133	137
20	3325	158

Table 1.2: Pounds of Force Needed to Transport a Fifteen-ton Load

Palmer's book is less about the business of railroads and more about technology incident to their construction and operation. The full title of Palmer's book sets the stage:

> *Description of a Railway on a New Principle With Observations on Those Hitherto Constructed. and a Table, Shewing the Comparative Amount of Resistance on Several Now In Use. Also an Illustration of a Newly Observed Fact Relating to the Friction of Axles, and a Description of an Improved Dynamometer, for Ascertaining the Resistance of Floating Vessels and Carriages Moving on Roads and Rail-Ways*

---

[8] In fact, Gray also reports on Wilkes' experiments using words identical to Cumming's but without quotation marks or a citation.

Palmer's view is that what is holding railroads back (as it were) is friction. Setting out to find sources of friction he encounters a counter-intuitive state of affairs, namely that freshly oiled axles offer greater resistance to the rotation of the wheels than axles that have been in use for some time. Palmer puzzles:

> During a succession of experiments, occasionally for many months, I invariably perceived, that when fresh oil was applied to the axles of the carriage, the resistance was increased, and it required the ordinary motion of the carriage for several days to restore the resistance to its usual standard. [185, p. 21]

In the process of changing various experimental conditions to see if he could effect the friction of the wheels against the axles he noticed that if "the axles were made perfectly clean, and then simply moistened with oil by the finger" the resistance was not increased. He goes on to discover why this is the case[9] and designs a new axle cross-section whose "resistance was one-tenth less than the former standard."

From his own experiments and those of others, Palmer concludes his discourse with the following practical guidance:

> [A] line of railway should be as straight as possible; it should be nicely adjusted to that plane which is most profitable; its parts should be as few as circumstances will admit; the touching surfaces should be hard and smooth, and should not be exposed to extraneous matter lying upon or adhering to them. [185, p. 29]

Noting that "it appeared desirable to elevate the surface from the reach of those obstacles" and that "[elevating] two lines of rail for the purpose of supporting a carriage, could not be accomplished at a sufficiently moderate expense" he "endeavoured to arrange the form of a carriage in such a manner that it would travel upon a single line of rail, without the possibility of overturning."

In case you haven't guessed it by now, the New Principle in the title of Palmer's book was the monorail.

1.4 The Practitioner's Experimental Essay

In the introduction to his book about how scientists learned to create, certify, and disseminate knowledge at the dawn of the scientific revolution, Peter Dear visualizes the history of experimental philosophy as follows:

---

[9]It was the quantity of oil applied to the axles that was the problem. An excess of oil functioned as a wedge at the point where the axle rested on the wheel.

> Boylean experimental philosophy was not the high road to modern exper-
> imentalism; it was a detour. [71, p. 3]

What Dear sees as a detour I see as a fork in the road. Dear took one fork—the scholarly fork—and found that it led back to the highway of laboratory-based experimentalism. Nineteenth-century practitioners took the other fork—the low road, if you must—experimentalism in the field, at scale, and for practical purposes. They, I will argue, were the inheritors of Boylean experimental philosophy.

Steven Shapin in his article "Pump and Circumstance: Robert Boyle's Literary Technology" examines the literature of experimental philosophy in its role as a meeting place between an author who has created a matter of fact and a reader who wishes to harness matters of fact for practical purposes. Shapin summarizes the Boylean manner of discourse the following way:

> If one wrote an experimental report in the correct way, the reader could
> take on trust that these things happened. Further, it would be as if that
> reader had been present at the proceedings. He would be recruited as
> a witness and be put in a position where he could validate experimen-
> tal phenomena as matters of fact. Therefore, attention to the writing of
> experimental reports was of equal importance to doing the experiments
> themselves. [209, p. 493]

An experimental report written in this manner and with this intent was first described by Robert Boyle. He called it an *experimental essay*.

Maurizio Gotti[10] has distilled from Boyle's writing and the style guides of the Royal Society the following distinguishing characteristics of an experimental essay:[11]

Brevity Sentences should be as concise as possible with no space given to unnecessary details.

Lack of Assertiveness There is no need for the author to arrive at definite conclusions or to systematise the results obtained.

Perspicuity Rhetorical embellishment should be avoided.

Simplicity Use of simple verb-forms and sentence constructions.

Objectivity Use of modal auxiliaries and verbs like *seem* and *appear* to show the author's uncertainty.

Boyle called this style of literary discourse a "naked way of writing." Joe Friday would say "Just the facts, m'am."

---

[10] See [117] and [118]

[11] See [157, p. 123].

The explicit intent of the experimental essay is to have the experimental outcomes speak for themselves. Sufficient detail must be given with regards to the experimental conditions and the experimental outcomes so that the reader is able to form an independent opinion as to the reliability and robustness of the connection between the two. In particular, the formation of this opinion is to be uncolored by any conjectures the author of the essay may harbor regarding cause and effect. The experimental essay is to be equivalent to the reader having been present at the performance of the experiment.[12] When written according to Boyle, the reader will walk away from an experimental essay muttering to himself, "Well, that's just how things are."

Starting with the same experimental data, the travelers on Dear's low road—the practitioners—engage in deduction while their scholarly friends on Dear's high road engage in induction. Whether you deduce or induce from this common starting point is determined by why you came to the experiment in the first place. As succinctly put by Edwin Layton:

> Basic science aims at knowing; it seeks generality and exactitude, even at the price of a good deal of idealization. But engineering science serves the needs of practice, even when this involves loss of generality and acceptance of approximate solutions. These differences in orientation go very deep. [162, p. 88]

and even more succinctly by William Whewell:

> Art and Science differ. The object of Science is Knowledge; the objects of Art, are Works. [250, p. xli]

Terry Reynolds goes on to characterize the difference between experimentation by scientists and experimentation by technologists in the following way:

> Science requires some system of conceptual or theoretical knowledge. Experiments, even quantitative experiments, are not scientific in and of themselves, unless they are directed by some type of conceptual structure....Thus, the use of models to test designs and ideas had deep roots in practical technology; it was not a gift from science. [198, pp. 284–285]

To the practitioner it is immaterial if the scientific consensus of the day is Newtonian or Einsteinian, quantum or string, as long as the river keeps turning the water wheel the way it always has.

---

[12]...a virtual witness to the experiment in some tellings.

1.4.1   Context and Continuity

As the vast literature of the history of science surely attests, the line between practical and scientific experimentation is not bright. The practitioner and the scholar both seek understanding and they both conduct experiments in pursuit of their quest. Where they differ is in the use they intend to put newly acquired understanding not the methods by which they to come to it. Because of this difference—and because Boyle was concerned primarily with scholarly understanding—I append two additional properties of an experimental essay to Boyle's to distinguish a scientist's experimental essay from what might be thought of as a practitioner's experimental essay:

> **Context** The experiment is conducted with apparatus and reported in terminology common to the readers's context.

> **Continuity** The experimental results are presented in a manner that can be readily adapted by the reader to conditions close to the experimental conditions.

The **context** property of a practitioner's experimental essay requires among other things that the experimental equipment is readily available for try-it-yourself purposes and that the units of measured quantities are familiar to the practitioner. Satisfying the context property also requires that the experiment be conducted at scale or nearly. Practitioners are often accused of being ignorant of the scientific and scholarly literature of the day. Doubtless, some were. But others who were familiar with the literature were skeptical of the findings of savants not because they thought that the findings were in any sense faulty but they knew too well that factors invisible or unaccounted for on the laboratory bench could be controlling in the field.

The **continuity** property of a practitioner's experimental essay requires that the author include a way to estimate what the experimental outcome would be for experimental conditions close to those reported in the essay. This could be a table that supports interpolation such as Gray's or a computational formula such as Wilkes'. Either way, the reader can plug in their own values to adapt the outcome to the problem at hand.

1.5   Kirk Boott's Practical Purpose

Getting back to the meeting of the directors of Merrimack Manufacturing, what may have led Boott to borrow Wood's *A Practical Treatise on Rail-Roads*? Rereading the quote taken from the introduction to Wood's book provides the answer:

> The greatest care has been used in the prosecution of the different experiments, and the most minute details are given, in order that the reader

may be able to judge of the credit to which they are entitled; our object has been to furnish practical data on the subject, and in doing so, not to assume any theory, or deduce any proposition, which is not supported by experiment. [252, p. vi]

There is little question that Wood was writing a practitioner's experimental essay. He promises a detailed report of all experimental conditions and outcomes and, furthermore, says that any discussion of the results of his experiments will stick very close to the data and will not venture into theories or hypotheses regarding causes.

Like Palmer whom he references,[13] Wood is particularly interested in the friction of carriage wheels against their axles. Here is his description of one of his experiments:

A perfectly straight plane of the Edge Railroad, with a uniform and regular inclination, was taken, the declivity of which was such as would cause the carriages to descend with an accelerated velocity; a carriage was placed upon it, and allowed to descend freely, and the space it passed over in successive portions of time, was marked with the utmost accuracy in the following manner—standing upon one end of the carriage, and aided by an assistant, at the end of every ten seconds I made a mark upon the plane where the carriage happened to be, and afterwards measured the distance between those marks, which gave the space passed over in each successive period. The carriage was first put in motion at the top of the plane, by a slight impulse, only sufficient to overcome its *vis inertia*. The descent of the plane was 1 yard in 200 yards, or 134 inches, in 13968 inches. [252, p. 198]

Wood is using readily available experimental apparatus, familiar units, and working at scale. Context requirement? Check.

Wood includes the following table of experimental results for the experimental condition "Loaded carriage, weighing 9,400 lbs., wheels 33 inches, axles 2 inches diameter."

---

[13] Wood is not, however, overly impressed with Palmer's skills as an experimental essayist: "Mr. Palmer, in the description of his Railroad, states the result of some experiments made on the friction of carriages moved along different kinds of Rail-roads. He makes the resistance considerably greater than Mr. Grimshaw, amounting to the eighty-seventh part of the weight, as found upon the Edge Rail-road, from the Penryn slate-quarries; but, as this must have been owing to some difference in the construction of the rail-way; and, as Mr. Palmer does not give any detail of his experiments, we are not, therefore, capable of judging of the cause of such an anamoly [*sic*]."

Descent Time (sec.)	Actual Space (ft.)	Calculated Space (ft.)
18	25.0	26.0
28	71.9	63.0
38	124.6	116.0
48	205.2	185.0
58	276.5	270.0
68	384.7	371.8
78	506.1	489.2
88	645.5	622.6
98	785.3	772.2[14]
108	939.6	937.9
118	1081.6	1119.6
128	1266.5	1318.3

Table 1.3: Wood's Table for Distance Traveled

If you were curious about the distance traveled after 63 seconds, halfway between 58 and 68 in Wood's table, you would take the average of the measured space at 58 seconds and the measured space at 68 seconds and make a new row in his table:

Descent Time (sec.)	Actual Space (ft.)
58	276.5
63	330.6
68	384.7

Table 1.4: Linear Interpolation in Wood's Table

Continuity requirement? Check.

---

[14]The value in the table printed in Wood's book is 722.2. The printer may have been speed-reading and seen a '7' rather than a '2'. Printers as opposed to authors were far and away the greatest source of errors in mathematical tables, big and small. An ill-selected sort in the middle of a word has a good chance of being caught by a proof-reader even if he didn't understand the text. Not so for a table of numbers. 722.2 is just as reasonable as 772.2 to a journeyman typesetter.

But Wood goes a step further in satisfying the continuity requirement; he gives the reader a simple formula for distance traveled that is a more compact and more useful summary of his experiment:

$$S = \frac{G - F}{W} \times rt^2 \qquad (1.2)$$

In Wood's formula, W is the weight of the carriage in pounds, 9,408 in his example; t is the elapsed time from the nudge sending the carriage down the hill; r acceleration of gravity in feet per second divided by 2, about 16.5; G is the weight of the carriage times the slope of the path or what Wood calls the *gravitating force of the body*, 9408 × $\frac{134}{13,968}$ or about 90.25; and, finally, F is the retarding force, the friction of the carriages, taken to be 44.4 from previous experiments. Wood's formula for the approximate distance of the carriages from the starting point after t seconds is the following:[15]

$$\frac{90.25 - 44.4}{9408} \times 16.5t^2 \approx 0.0805t^2$$

For $t = 63$, Wood's formula gives 319.5. Linear interpolation in Wood's table for $t = 63$, gave 330.6, an error of about 3.5%.

The entries in the right-most column of Wood's table, Table 1.3, were computed using his formula. The estimated values are displayed right next to the measured values so that the readers can judge the reliability of the formula for themselves Continuity requirement? Double check.

Finally, in compliance with Boyle's urging that unexpected outcomes and experimental failures should be reported too, Wood calls the reader's attention to the reversal of the relationship between the actual and computed values in the bottom two rows in the table. He notes in closing the discussion of his experiments:

> It will be seen, from the preceding experiments, that the actual passed over is greater, until a certain period of the time, than the calculated space; which arose from the wind blowing pretty strong in the same direction of the line of the plane, as that in which the carriage was descending, which had the effect of urging it forward, until its velocity became equal to that of the wind. [252, p. 199]

Boott and Jackson were going to have to draft a petition to the Massachusetts legislature to grant articles of incorporation for their railroad project and they fully expected their petition would be opposed by proprietors of the Middlesex Canal.[16] Their petition would have to be firmly founded on matters of fact because matters of

---

[15] The symbol $\approx$ is the mathematical notation for 'approximately.'

[16] The Proprietors of the Middlesex Canal did indeed file a remonstrance on February 13, 1830,

fact are harder to challenge than abstract theories of public good and the social benefits of improved internal transportation. Undeniable facts about the relative merits of
canals versus railroads based on experiments conducted in the field and at scale were
what Kirk Boott needed and they were what the books he borrowed delivered.

A primary contention of the prospective proprietors of the Boston and Lowell railroad set forth in their petition to the legislature was that the transportation of goods by
rail was more fuel efficient than transporting them by canal[17] and therefore the lower
operating expense of the railroad would benefit everyone except, perhaps, the proprietors of the Middlesex Canal. How many bales of hay or bushels of coal one had to
expend to move a given load of goods a given distance was the top contributor to the
cost of transport and this cost figure turned on the resistance of the loaded carriage at
rest to being put in motion, be it a barge on a canal, a wagon on a road, or a car on
rails.

Both Palmer and Wood devote considerable attention to computing and comparing the power of horses and steam engines when applied to moving loads on canals
and railways so there is no mystery about why Boott might have given their texts a
close read. Wood summarizes his book in a simple statement which surely would
have caught Boott's attention. "We derive the conclusion," Wood writes…

> …that so long as the expence [*sic*] of one loco-motive engine does not ex-
> ceed that of four horses, and their attendants; then goods can be conveyed
> with the same expenditure of motive power at six miles an hour upon a
> Rail-road, that they can be conveyed at two miles an hour upon a canal.
> [252, p. 311]

If the first 300 pages of Wood's book convinced Boott that he, Wood, was telling it
like it is, then here is an experimental finding that Boott could take to the state house
and then, the legislature willing, to the bank.

Such quantitative rules-of-thumb are occasionally called rules, theorems or propo-
sitions in practitioner experimental essays. They are not to be confused with the laws
found in scientific experimental essays. They do not stake any claim to insight into
underlying causes, phenomena, or mechanisms that might connect experimental conditions to experimental outcomes. Nor do they claim relevance to other settings. They
simply summarize a collection of experimental outcomes in a particular circumstance
in a handy rule-of-thumb.

[76, pp. 25–31] but to no effect. The incorporation of the Boston and Lowell Rail Road Corporation was
approved by the Massachusetts General Court on June 5, 1830, with John F. Loring, Lemuel Pope, Isaac P.
Davis, Kirk Boott, Patrick Tracy Jackson, George W. Lyman, and Daniel P. Parker as proprietors. Writing
thirteen years later, Caleb Eddy, a director of the Middlesex Canal Corporation vented: "But it was thought
best to plunge the knife, to the hilt, into the existing rights of the canal proprietors." [76, p. 18] The Middlesex Canal Corporation went belly up on April 4, 1860, but not before Eddy tried to shape-shift the canal
into an aqueduct and sell it to the City of Boston as means for supplying water to the city.

[17]…not to mention that canals couldn't be used in winter.

1.6   Purposeful Reading in Experimental Philosophy

Patrons of the Athenæum were active participants in Boston's many scientific and technical communities as well as busy promoters of industrial projects of every nature: mills, railroads, mines, foundries, and canals to name just a few of the latter. They were founding members of scholarly societies such as Nathaniel Bowditch and George B. Emerson. They were promoters of economic development such as Thomas Handasyd Perkins and Patrick Tracy Jackson. They were physicians such as Charles T. Jackson and Augustus A. Gould. They were industrialists such as Cyrus Alger and Erastus B. Bigelow. They were inventors such as Daniel Treadwell and Paul Moody. And they were architectural engineers such as Alexander Parris and George M. Dexter. They spent their days working, reading, writing, corresponding, planning, and building at the intersection of science, art, and commerce. The books and journals they borrowed at the Boston Athenæum contained up-to-date information on technological innovation, reports of large engineering projects, news on patents, as well as the latest findings of academic researchers.

In the next chapter I describe the data set used for this study; namely, the contents of the first four volumes of the Boston Athenæum's books borrowed register covering the years 1827 to 1850. In addition to a little background history about the borrowing of books at the Athenæum, I give some overall statistics of the entries in the registers. Finally, I describe a graphic display that will be employed throughout the book to visualize the ebb and flow of interest in various scientific and technical subjects.

In the third and fourth chapters, I focus on books about engineering, particularly those dealing with harnessing the power of water and steam. In the period covered by the Athenæum's first four books borrowed registers, the second quarter of the nineteenth century, water wheels were slowly giving way to hydraulic turbines as a way to capture power from New England's rivers and to steam power as a prime mover. Steam was also starting to be used in deference to open fires as a means of heating interior spaces.

In the fifth and sixth chapters, I shift attention to titles of a more scientific nature. Chemistry, geology, mineralogy, and botany were all popular topics, both in and of themselves and as they connected to practical purposes such as farming, mining, and *materia medica*.

Mathematical instruments are central to all experimentation so it is not unexpected that mathematical models and mathematical reasoning were important parts of the practitioner's experimental essay particularly in meeting the continuity condition. In Chapter 7, I consider reading in mathematics *per se* in order to explore the kind of mathematics that Athenæum readers sought and were familiar with in their reading. Mathematics plays an outsized role in astronomy and architectural engineering texts so reading in these two fields is also covered in this chapter.

In Chapter 8 I consider the patents held by Athenæum readers. A patent application is a practitioner's experimental essay codified in law. In order to be granted a patent the petitioner is obliged to describe his invention in sufficient detail that it can be practiced by anyone skilled in the art when the patent expires. Twenty-four of the readers in experimental philosophy were granted a U.S. patent, many of them more than one. As many of the patent applications were written by the inventor himself and often in his own hand, these documents are in a very real sense halfway between what they read and what they built.

The appearance of experimental philosophy in children's literature and in Boston's public school curriculum is the subject of Chapter 9. Chapter 10 finishes with some closing remarks.

Throughout the following chapters I take a very broad view of the Boylean experimental essay, undoubtedly broader than he intended. It is a view of learning through experimentation and discovery that has as much in common with Maria Montessori and Johann Pestalozzi as it does with Michael Faraday and Antoine Lavoisier; namely, experimentation and discovery for practical purposes.

Chapter 2

The Library of the Boston Athenæum

2.1 Solomon Miles and Tables of Logarithms

On November 6, 1827, Solomon Miles borrowed *Tables Portatives de Logarithmes* by François Callet as well as Charles Hutton's *Mathematical Tables.* Both books were due back on December 6. What do these charges tell us, if anything, about Solomon Miles and his works?

In the first half of the nineteenth century the library of the Boston Athenæum was one of the largest by number of volumes in America, being surpassed only by the collections at Harvard and the Library Company of Philadelphia. As witnessed by the Athenæum's 339-page catalog of 1827, its holdings in science and technology were broad and deep. There were seventy-four volumes of the *Annales de Chimie* and fourteen titles by Johannes Kepler as well as a first edition of Boyle's *Experimentorum-Physieo-Mechanieorum* [39].

At the time he borrowed the mathematical tables of Callet and Hutton, Solomon Miles was head master of Boston English High School. The sub-master during much of his tenure was his long-time friend, Thomas Sherwin. In 1830, three years after Miles took home the tables of Callet and Hutton, Miles and Sherwin published *Mathematical Tables; Comprising Logarithms of Numbers, Logarithmic Sines, Tangents, and Secants, Natural Sines, Meridional Parts, Difference of Latitude And Departure, Astronomical Refractions.* In the Advertisement to their book, the authors note:

> The Tables, comprised in this volume, have been very carefully compared with the best English and French Tables; and they will be found, it is

believed, not inferior, in point of correctness, to any similar Tables in use. [176, p. v]

Miles and Sherwin didn't say—as was *de rigueur* when publishing mathematical tables—exactly which English and French tables they read their tables against but based on Miles' charges in 1827 it is safe bet that the tables of Callet and Hutton were among them.

2.2   Purposeful Reading at the Boston Athenæum

Neither a library nor a museum but rather a quiet place in the middle of the city to catch up on the news, chat with your colleagues, browse the latest books, explore a cabinet of curiosities, and perhaps, on an odd day, conduct a chemistry experiment or two; a combination reading room, meeting place, reference desk, laboratory, and lecture venue for all intellectual pursuits. Such was and still is the practical purpose of the Boston Athenæum.

The first athenaeum, built by Hadrian in Rome in 123,[1] was a multi-level complex that included a 900-seat auditorium with an 36-foot high ceiling where philosophers, politicians, and poets could hold forth on works in progress and topics of the day.

The initial version of an athenæum for Boston was somewhat more modest: a rented suite of rooms in Joy's Building on Congress Street named grandiloquently the Anthology Reading-Room and Library. The vision of the promoters of Boston's nascent athenæum were no different than Hadrian's however. As described in the prospectus soliciting support for the fledgling institution:

> The Reading-room shall be furnished with seats, tables, paper, pens, and ink; with the Boston papers, and all the celebrated gazettes published in any part of the United States; with the most interesting literary and po-litical pamphlets in Europe and America; with magazines, reviews, and scientific journals; London and Paris newspapers; Steel's Army and Navy Lists; Naval Chronicle; London and Paris booksellers' Catalogues; Parlia-mentary Debates; bibliographical works, &c. &c. [197, p. 7]

The announcement of the opening of the Anthology Reading-Room and Library included rules of proper conduct among which was the following:

> [N]o book, pamphlet, or newspaper is ever to be permitted to be taken from the rooms by subscribers; so that the patrons of the institution may be certain at all times of finding any publications, which they may have occasion to read or refer to. [197, p. 12]

---

[1] ...and rediscovered in 2009 in during excavations for a new subway station.

The reading room was just that, a reading room. It was not a circulating library, at least not yet.

The response to the solicitation for subscriptions to the Reading-Room was overwhelming and very quickly gave way to a more grandiose plan. But rules are rules and it would still be a reading room and not a lending library. The *Memoir of the Boston Athenæum* published in May of 1807 soliciting subscribers to a new and improved reading room did give just a little bit of ground on the matter of borrowing however:

> None of the papers or periodical works are to be taken from the rooms, except in the case of the indisposition of any proprietor or subscriber, who may have the use of the newspapers at his house, at some convenient time after their arrival, under such regulations, as shall be prescribed. Duplicates are to be provided of all those books permitted to circulate, it being intended that one copy of every work belonging to the Library shall always remain in it; so that the proprietors and visitors of the Athenæum may be certain at all times of finding any work, which they may have occasion to read or consult. [210, p. 5]

When the reading room, now named the Boston Athenæum, reopened its doors on August 11, 1808, the book collection was again locked down tight:

> All new books, pamphlets, magazines, and reviews shall be placed on the tables of the library or reading-room, and remain there for such length of time as may appear necessary for their perusal; and no book, pamphlet, review, magazine, or newspaper shall be taken by any proprietor, subscriber, or visiter, out of the rooms. Any proprietor or life subscriber, infringing this article, shall forfeit his privilege in the institution for one year; and annual subscribers shall forfeit the privilege of their subscription by a similar offence. [197, p. 49]

Five departments were described in the Athenæum's Articles of Incorporation, one of which was…

> …a Laboratory, and an Apparatus for experiments in chemistry and natural philosophy, for astronomical observations, and geographical improvements, to be used under the direction of the corporation. [210, pp. 5–6]

To this end, and shortly after the enterprise was up and running, the Athenæum took possession of a portion of the scientific apparatus and mathematical instruments belonging to the Society for the Study of Natural Philosophy.[2] The Society was in the throes of disbanding and was looking for a final resting place for its assets.

---

[2] Not to be confused with the Boston Society of Natural History.

Roughly a decade later, in April of 1817, the American Academy of Arts and Sciences deposited its library of books with the Athenæum. In May of the following year this arrangement was made permanent with the proviso that members of the Academy could take home books from the Academy's collection but not from the Athenæum's collection. Athenæum members could use the Academy's holdings on the premises but they were not permitted to leave with them. At about the same time the books of the Massachusetts Agricultural Society were also deposited with the Athenæum. The Trustees of the Athenæum were surely becoming aware of the administrative complexities these depository relationships were generating but did nothing proactive at the time to address them.

An updated set of by-laws of the Boston Athenæum were adopted in February of 1822 as the enterprise prepared to relocate to James Perkins' expansive fourteen-room mansion down on Pearl Street. As part of the update, some of the rules and regulations regarding proper conduct on the premises were modified to reflect the physical circumstances of the new location but not the rule on borrowing books:

> No book or other article to be carried out of the rooms on any pretext. Offenders, if proprietors, forfeit their rights for one year; if annual subscribers, their subscription. [197, p. 77]

There were no explicit carve-outs for members of the American Academy or the Agricultural Society in the new rules but in January of 1824 the agreement with the Academy was renewed on the same terms as the 1818 agreement with respect to the borrowing of books.

The Church of England's provincial library at King's Chapel was one of over seventy libraries created and nurtured by the the Church of England's Commissary to the New World, Thomas Bray. [159] The founding tranche of over 200 volumes arrived in Boston in 1698. In 1807, the collection was deposited with the Fourth Social Library of Boston, the Theological Library, and then in 1823, when the Theological Library was absorbed the Athenæum, the collection was moved to Pearl Street. It is not known what the borrowing rules were for the King's Chapel collection either before or after the merger but what is certain is that the Theological Library would not be the last social library that would be consolidated with the Boston Athenæum so perhaps a word about these institutions would not be out of place.

2.2.1   Boston's Social Libraries

A Social Library was an institution recognized and chartered by the State of Massachusetts in accordance with an act passed in 1797. The act defines a Social Library as...

...any seven or more persons in any town or district, who have or shall become Proprietors in common of any library, to form themselves into a society, or body politic, for the purpose of holding and using such libraries.

The law was amended a number of times to make explicit its application to specific kinds of social libraries; in 1806 to create Military Social Libraries, for instance. Charles Bolton in his article *Social Libraries in Boston* [19, p. 333] lists four social libraries that were up and running in Boston at the dawn of the nineteenth century:

- The First Social, or the Social Law Library, 1803
- The Second Social, or the Boston Medical Library, 1805
- The Third Social, or the Scientific Library, 18—
- The Fourth Social, or the Theological Library, 1807

The dashes at the end of the entry for the Scientific Library are Bolton's. Concerning this library Bolton says "This third social library is still somewhat of a mystery, for we find no contemporary organization. Twenty years later came the Massachusetts Scientific Library Association." A fifth social library not on Bolton's list was constituted in 1809, the Social Architectural Library, also known as the Architectural Library of Boston. [174, p. 83]

A social library much like a web site is easier to start than to keep going. Figure 2.1, adapted from [173], illustrates the rapid growth and slow decline of American's social libraries. The dark line is the number of social libraries within New England and the light line is the number of social libraries outside of New England.

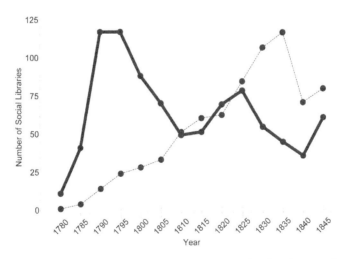

Figure 2.1: Social Libraries within New England and Beyond

When enthusiasm walks out the door the members left behind are faced with the problem of disposing of the books. In Boston that usually meant heading down to Pearl Street and initiating merger talks with the Athenæum. The Boston Medical Library and the Theological Library both made the trip. While there were merger discussions with the Architectural Library, this library was eventually absorbed by the Massachusetts Charitable Mechanic Association.[3] [174, p. 85]

The idea of a scientific social library was refloated in January of 1826 by Athenæum reader, Edward Brooks. The committee formed to solicit subscriptions consisted of Brooks together with Israel Thorndike, Jr., Amos Lawrence, John A. Lowell, George B. Emerson, John C. Gray, John Lowell, Jr., William Sturgis, Daniel Treadwell, and Enoch Hale. The bill to incorporate the second scientific library passed the Massachusetts legislature as part of a package of bills including one that abolished whipping as a punishment in the Commonwealth. The committee went right to work soliciting subscriptions and must have met with some success because two months later, on Saturday, March 18, 1826, an organizing meeting was held at the Athenæum. Sturgis was elected President and a number of our readers including Warren Colburn, Jacob Bigelow, Daniel Treadwell, George B. Emerson, Abbot Lawrence, Francis Cabot Lowell, John Ware, Joseph Coolidge, and Solomon Willard were elected trustees.

2.2.2 The Bowditch Committee

In 1826, the Trustees of the Boston Athenæum finally decided to bring clarity and order to an increasingly complicated tangle of operational and ownership issues surrounding the borrowing of books and the absorbing of circulating libraries. The Athenæum's borrowing policy, rooted as it was in the operation of a reading room, was the bone in the throat that was going to have to be extracted. Recently elected a trustee of the Boston Athenæum and Boston's antebellum organization man,[4] Nathaniel Bowditch, was given the task:

> On the 18th of March, 1826, a special meeting of the Proprietors was held at the call of the committee appointed on the subject of the union of libraries and the circulation of books, of which Nathaniel Bowditch was chairman. [197, p. 95]

The singular event that called the Athenæum's trustees to action and triggered the creation of the Bowditch committee was the Boston Medical Library knocking on the

---

[3]The Architectural Library was by far the smallest of the five social libraries, never exceeding a couple hundred volumes. The book collection was stored in Alexander Parris' office before it was finally remaindered to the Mechanic Association. As Parris was member of the Boston Athenæum it is possible that one or two architectural books that Parris donated to the Athenæum were from the Architectural Library. Books that he had initially contributed to the Architectural Library are obvious candidates.

[4]...and not incidentally American's first internationally recognized mathematician.

door, books in hand. As told from the point of view of the Medical Library it was not a happy day:

> The Athenæum had already absorbed many societies, and in 1826 the Trustees appointed a Committee…to consider the expediency of uniting in the Athenæum the principal circulating libraries of Boston,[5] that the deficiencies of the larger library might be overcome. The Medical Library seems to have fallen victim to this Committee. [100, p. 18]

Bowditch was not a physician but he was a respected member of the Massachusetts medical community at large. He had been consulting with the Massachusetts General Hospital on annuity matters since 1816 and was also a member of the Salem town committee that helped raise funds for the hospital. He moved to Boston in 1823 to become the full-time corporate actuary of the Massachusetts Hospital Life Insurance Company.[6]

As if dealing with mergers and the touchy topic of borrowing of books weren't enough, Bowditch's committee was also charged with reorganizing the book shelves, creating and printing a catalog, completing broken sets, and correcting any other imperfections in the library might uncover.

In the event, the Boston Medical Library became part of the Boston Athenæum on May 26, 1826, and its two thousand plus books were added to the Athenæum's collection. Thirty of the thirty-one proprietors of the Boston Medical Library became proprietors of the Boston Athenæum, being issued shares 206 through 235. The odd man out was Blowers Danforth.[7] Danforth did, however, elect to became a regular member of the Athenæum with book borrowing privileges.

By July of 1826, the exact date is uncertain, the Massachusetts Scientific Library Association was folded into the Athenæum. As reported in the *United States Literary Gazette*:

> BOSTON ATHENÆUM. The value and usefulness of this excellent institution have within a few months been greatly increased by a union with the Medical Library and the Scientific association, as well as by a munificent subscription in aid of its funds.…With [these] additions to its former means, the Boston Athenæum will become far superior to any institution of the kind in the United States.[8]

---

[5]It is probably closer to the truth to say that the Athenæum absorbed the principal *social* libraries of the day rather than the principal circulating libraries. One might also quibble with 'principal.'

[6]The company had been chartered in 1818 but didn't start conducting business until 1823. Athenæum benefactors, James and Thomas Handasyd Perkins, were among the founding directors. One of Bowditch's lesser known publications, *Proposals of the Massachusetts Hospital Life Insurance Company* [32], is one of the earliest actuarial tracts published in America.

[7]…or Bowers Danforth as he became known after he changed his name on June 23, 1831.

[8]Volume IV, July 1826, p. 310, as reported in the *Boston Daily American Statesman* of July 24, 1826.

Commenting on the reconstitution of the Massachusetts Scientific Library Association as the scientific department of the Athenæum, the article goes on to say:

> This department will also be rendered much more complete, by the sum subscribed in February for completing the Transactions of the Royal Societies and Academies of Sciences in London, Edinburgh, Dublin, Paris, Petersburg, Berlin, Turin, Göttingen, Stockholm, Madrid, Copenhagen, and Lisbon making in the whole, one of the most complete scientific libraries in the United States.

Getting back to the Bowditch committee, its final report regarding borrowing books was delivered on December 13, 1826. It read in part as follows:

> The Committee to whom was referred the terms & restrictions under which books should be taken out, with power to make rules concerning the same, beg leave to Report, that they have considered the following rules taken chiefly from the regulations of similar institutions as made suited to the purposes of the Athenæum
>
> By-laws relative to taking
> Books from the Boston Athenaeum.
>
> 1.   Proprietors of Shares & Life share holders, by paying an annual assessment of <u>Five dollars</u> shall have a right to take out books from the Athenæum. Also, such subscribers to the Scientific Library as have already complied with the conditions of the transfer of the Library to the Athenæum, & paid the first assessment, shall by paying an annual Assessment of <u>Ten Dollars</u> have the right to take out Books in the same manner as the Proprietors, Provided however, that this right shall cease upon failure to pay this annual assessment & that no person shall have a right to take out any book title [unless] assessments & other dues are paid.
>
> 2. Any person entitled to take out Books under the last article, may have at any one time, three Volumes.
>
> 3. Books may be kept out of the Library, one calendar month & no longer, and every person shall be subjected to a fine of twenty cents a week for every volume retained beyond that time.
>
> 4. Every book shall be returned in good order, regard being had to the necessary wear of the book with good usage. And if any book shall be lost or injured the person to whom it stands charged shall replace it by a new volume or set, if it belonged to a set, or pay the current price of the

volume or set to the Librarian; and, therefore, the remainder of the set, if the Volume belonged to a set, shall be delivered to the person so paying for same.

5. All Books shall be returned to the Library for examination on the second Wednesday of May annually and remain one fortnight.[9] And every person then having one or more books and neglecting to return the same, as herein required, shall forfeit & pay a fine of one dollar.

6. When a written request shall be left with the Librarian for a particular book then out, it shall be the duty of the Librarian to retain the same, for the person requesting it, for one day after it shall have been first returned.

7. The Librarian will attend for the delivery & return of Books from noon till II o'clock every day, Sundays excepted.

8. The Trustees may, on special occasion, permit any person to use the Books belonging to the Athenæum, under such restrictions as they may think proper to impose.

9. Very rare or costly Books, which cannot easily be replaced, are not to be taken from the Athenæum, except by a vote of the Trustees; and all new periodical Publications and new Works imported from Europe shall be withheld from circulation; so long as the Trustees may deem expedient.

All of this is an overly detailed way of explaining why the first volume of the Boston Athenæum's books borrowed register is dated twenty years after the Athenæum's founding.

What makes this sketch of these years relevant to this study, however, is that it chronicles the absorption into the Boston Athenæum of a considerable number of readers and books centered on science and technology. It was in part by merging with the Scientific Library and the Medical Library and by hosting the library of the American Academy of Arts and Sciences that the the Athenæum became, in the words of the *Literary Gazette*, "far superior to any institution of the kind in the United States."

2.2.3   The Books Borrowed Registers

In order to borrow a book from the Athenæum's collection, you carried the book you wished to take home to the circulation desk. There, the books borrowed register would be opened to your personal page and the following data items would be written from left to right on the next blank line:

---

[9]The purpose for this by-law was so that the collection could be inventoried. It would be thirty years in the future that the sixth librarian of the Athenæum, William Lane, would invent the due date card.

Field 1. the date the book was being borrowed

Field 2. the number of the shelf on which the book was stored

Field 3. a telegraphic notation of the book's author and title

Field 4. ancillary data such as book size, edition, or volume number

Field 5. the date the book was to be returned

Figure 2.2 is an entry on page 17 of Volume I that records James T. Austin borrowing of Volumes 1 and 2 of Mill's *History of Chivalry* on June 7, 1828. The book was to be brought back by June 27, 1828, at which time it would be returned to shelf 933. Figure 2.3 is the entire page on which this charge appears. The charge in Figure 2.2 is about one quarter of the way down in the right-hand column of Figure 2.3.

Figure 2.2: James T. Austin's Charge of Mill's *History of Chivalry*

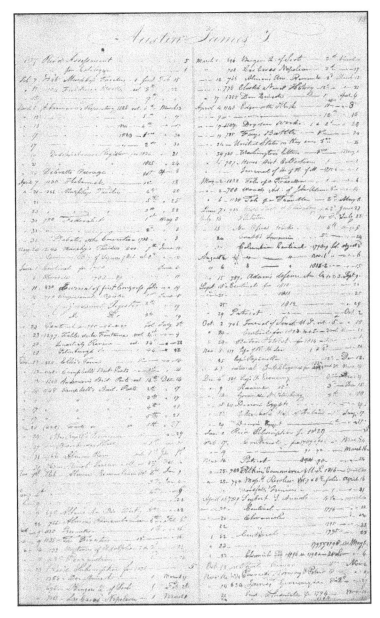

Figure 2.3: James T. Austin's Page in Volume I

When Volume I of the Athenæum's books borrowed register was initialized, a page was allocated to each name in the current membership roster in alphabetic order by last name and the name was written at the top of the register page. When a new member of the Athenæum elected borrowing privileges they were allocated the next unused page at the back of the volume. An unused page was also assigned to a reader whose charges overflowed their current page. When the register ran out of unused pages at the back, the pages of individuals who were no longer members or who had never borrowed a book were reassigned. Needless to say, the careful alphabetic ordering of the pages soon broke down.

Subsequent volumes may have followed the same initialization regimen although the first name in Volume II, "Parkman George MD," and Volume III, "W.P. Atkinson," are not alphabetically the first of all names in the respective volumes so the page assignment algorithm either changed or was less rigorously followed. It is also possible that if a reader's page filled near the end of a volume's lifetime, the continuation page for the reader would be allocated at the front of the next volume.

Figure 2.4 shows a page from Volume I that has been reassigned. The page was originally allocated to Edward Appleton and his 1827 charges were recorded in the left-hand column. Appleton must have not renewed his borrowing privileges because in 1834 the page was reassigned to Dr. Samuel G. Howe. Howe's charges can be seen written in the right-hand column.

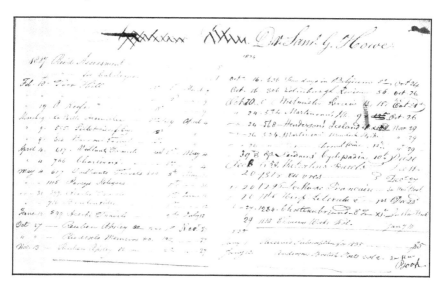

Figure 2.4: Register Page Assigned to Two Readers

When a register page contains the charges of more than one borrower it is almost always possible to determine which charges go with which name. One of the more curious examples of page reassignment is shown in Figure 2.5, page 86 in Volume III. It appears that George Francis Parkman, Jr., was assigned his father's page sometime between May 22, 1844, the first charge on the page and August 4, 1845, the last charge on the page. The name at the top of pages assigned to a George Parkman in Volumes I and II appears variously as "Parkman George MD," "Parkman George," or "Parkman Geo." In Volume III and IV the son's name appears uniformly as "George F Parkman."

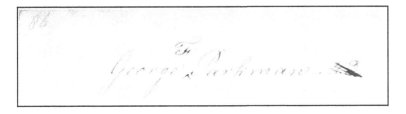

Figure 2.5: Reassignment of George Parkman's Page in Volume III

Finally, there are entries in each of the register volumes that spill over into next volume, the volume containing charges for the year after the first volume's closing year. Volume I, for example, nominally spans 1827 to 1834 but looking back to the lower right-hand corner of Figure 2.4 we see that Dr. Howe's subscription payment for 1835 on January 1, 1835, as well as his charge of Anderson's *Works of British Poets* on January 10, 1835, are recorded on his page in Volume I of the register.

In these cases, the charge is counted in the following volume. When a charge appears in both volumes, as is occasionally the case, it is only counted in the volume of its charge date. Every effort has been made to ensure each charge has been counted only once and counted in its proper year. That said, none of these counts should be taken to be exact.

Finally, there were four pages missing in Volume I, pages 94/95 and 226/227, and six pages missing in Volume II, pages 14/15, 176/177, and 178/179. Obviously, charges on these pages are not included in the study.

2.2.3.1   Counts of All Readers

Table 2.1 contains the counts of readers and charges in the first four volumes of Boston Athenæum's books borrowed registers. The counts in the column headed **Names** reflect the number of unique names appearing at the top of each volume's pages. A name that appears but was crossed out when the page was reassigned qualifies as a name for the purpose of this count as does a name that appears on an otherwise blank page. A name that appears on more than one page—crossed out or not—is counted only once. The counts in the column headed **Readers** reflect the number of individuals who charged one or more books.

The counts in the **Total** row for the **Names** and **Readers** columns reflect unique names across all four volumes. Since many names appear in more than one volume, the **Total** counts for these two columns are less that the sum of the counts for individual volumes.

Volume	Years	Pages	Names	Readers	Charges
I	1827–1834	533	340	302	36,987
II	1835–1843	543	324	321	47,915
III	1844–1849	467	360	334	24,995
IV[11]	1849–1850	552	377	343	12,151
Total	1827–1850	2,097	754[12]	703[12]	122,048

Table 2.1: Pages, Names, Readers, and Charges by Volume

Table 2.2 lists the total number of charges and the top reader for each year. Daniel Hammond was the most avid reader across all four volumes, logging 1,656 charges in the 7,955 days between his first charge on January 7, 1829, of Washington Irving's *Bracebridge Hall* and his last charge on October 19, 1850 of Sharon Turner's *History of England*. By raw count, Charles Folsom actually registered the greatest number of charges for each year from 1846 to 1850 but as he was the Athenæum's librarian at the time his charges undoubtedly reflect books borrowed on his own behalf as well as the charges made on behalf of "indisposed" patrons. Therefore, his name does not appear in Table 2.2.

---

[11] Volume IV includes 1,195 charges for 1851.
[12] Unique names across all four volumes.

Year	Total Charges	Top Reader	Reader Charges
1827	4,024	Danforth, Blowers	130
1828	3,655	Fairbanks, Gerry	89
1829	4,016	Fairbanks, Gerry	112
1830	5,033	Osgood, David	124
1831	4,877	Wigglesworth, Edward	168
1832	5,515	Wigglesworth, Edward	205
1833	4,582	Tuckerman, Henry Harris	184
1834	4,826	Tuckerman, Henry Harris	216
1835	5,378	Loring, Elijah	158
1836	5,288	Ellis, David	157
1837	5,269	Ellis, David	195
1838	5,082	Parkman, George Francis	163
1839	5,636	Tuckerman, Henry Harris	133
1840	5,010	Parkman, George Francis	139
1841	4,625	Hall, Joseph	112
1842	5,122	Bethune, John McLean	146
1843	4,700	Bethune, John McLean	119
1844	4,444	Ward, Thomas Wren	112
1845	4,579	Parkman, George Francis	103
1846	4,862	Burley, Susan	91
1847	5,466	Chapman, M.W., Mrs.	140
1848	5,610	Chapman, M.W., Mrs.	143
1849	5,489	Burley, Susan	205
1850	7,793	Burley, Susan	175
1851	1,195	Alexander, Andrew	38

Table 2.2: Total Charges and Top Reader by Year

2.2.3.2 Counts of Readers in Experimental Philosophy

Three hundred and eighty-nine readers recorded at least one charge of a title in experimental philosophy in the first four volumes of the Boston Athenæum's books borrowed register. These charges spanned five hundred and forty-two titles by three hundred and sixty-eight authors. Table 2.3 summarizes the number of readers and the number of charges in experimental philosophy for each volume and across all four volumes.

Volume	I	II	III	IV	I-IV
Readers	182	199	143	90	388
Charges	1,416	1,548	591	206 [13]	3,761 [13]

Table 2.3: Readers and Charges in Experimental Philosophy by Volume

There is a distinct fall-off in the borrowing of experimental philosophy titles starting in the middle of Volume II, say around 1840. This may be due to a shift away from science and technology in the Athenæum's book acquisition policy, the opening of competing libraries in and around Boston, or a simply a decline in the number of members interested in scientific and technical topics.

2.2.3.3   Subject Matter Categories

Each title in science and technology[14] included in the study has been classified into one of fourteen subject matter categories. The subject matter categories along with counts of the number of titles, readers, and charges in each are listed in Table 2.4. The titles in each category include both books and periodicals. Thus, for example, charges of Robertson's *London Journal of Mechanics* are included in the Mechanics category. The row labeled **Periodicals** in this table contains counts for these subject matter specific journals as a group unto itself.

---

[13] Includes twenty charges for 1851 recorded in Volume IV.
[14] ...with three exceptions noted below.

Category	Titles	Readers	Charges
Mechanics	90	136	481
Chemistry	52	130	459
Geology	45	172	445
Botany	71	106	335
Zoology	40	106	323
Astronomy	44	124	275
Mathematics	86	82	227
Entomology	25	89	222
Hydraulics	24	64	161
Pyronomics	10	65	130
Mineralogy	18	59	123
Ornithology	18	67	122
Conchology	12	44	114
Patents	6	59	108
Periodicals	13	89	315

Table 2.4: Titles, Readers and Charges by Subject

In addition to the fourteen subject matter categories there are three special categories that each contain a single title. Two of these are of periodicals, Jones' *Franklin Journal* and Silliman's *American Journal of Science,* and the third is Jacob Bigelow's *Elements of Technology.* Counts of readers and charges for these three special categories are listed in Table 2.5.

Author/Editor	Short Title	Readers	Charges
Silliman	*American Journal of Science*	38	98
Jones	*Franklin Journal*	38	92
Bigelow	*Elements of Technology*	34	46

Table 2.5: Charges for Special Categories

Overall, there were fifteen periodicals in the study counting thirteen subject matter specific periodicals and the two periodicals in Table 2.5. As a group, periodicals were charged 505 times by 123 readers. The number of readers and the number of charges for each periodical are given in Table 2.6.

Editor	Short Title	Readers	Charges
Robertson	*London Mechanics Magazine*	37	103
Silliman	*American Journal of Science*	38	98
Jones	*Franklin Journal*	38	92
Newton	*Repertory of Patent Inventions*	27	60
Lavoisier	*Annales de Chimie*	20	52
Curtis	*Botanical Magazine*	13	50
Greenough	*Transactions Geological Society*	9	26
	Transactions of the Linnean Society	5	10
Charlesworth	*Journal of Zoology*	1	5
Sowerby	*The Zoological Journal*	1	2
Treuttel	*Archives des Inventions*	2	2
Pritchard	*English Patents Granted*	2	2
Harlan	*Trans. of the Geological Society of Penn.*	1	1
Bruce	*American Mineralogical Journal*	1	1
	Trans. Royal Entomological Society	1	1

Table 2.6: Readers and Charges for Periodicals

Table 2.7 lists the readers with twenty or more charges in experimental philosophy across all categories and Table 2.8 lists the top reader within each subject matter category.

Reader	Charges	Reader	Charges
Bigelow, Jacob	86	Jackson, Patrick Tracy	30
Tuckerman, Edward	83	Randall, John	30
Emerson, George B.	78	Parsons, Theophilus	30
Parris, Alexander	72	Bates, George	30
Hall, Joseph	66	Reynolds, Jr., Edward	30
Dexter, Franklin	65	Burley, Susan	29
Lyman, William	61	Jackson, Charles T.	27
Martin, Enoch	61	Jones, Anna P.	26
Jenks, William	57	Guardenier, John	26
Heard, Jr., John	54	Phillips, Jonathan	26
Eddy, Robert H.	50	Warren, John C.	26
Nichols, Benjamin R.	49	Gould, Augustus A.	25
Goodrich, Samuel	48	Cruft, Edward	25
Belknap, John	47	Hooper, Samuel	24
Folsom, Charles	47	Warren, A., Mrs.	24
Sherwin, Thomas	45	Hale, Nathan	24
Gray, Horace	45	Harris, Thaddeus M.	24
Codman, Henry	40	Perkins, Stephen H.	23
Quincy, Jr., Josiah	37	Inches, Henderson	23
Lawrence, William R.	37	Treadwell, Daniel	23
Boott, John W.	37	Cooke, Sr., Josiah	22
Wyman, Rufus	36	Hooper, Henry N.	22
Capen, Nahum	36	Boyden, Uriah A.	21
Grigg, William	34	Gay, Martin	21
Miles, Solomon	34	Ware, John	21
Wigglesworth, Edward	33	Parsons, William	20
Scholfield, Arthur	33	Coffin, Joy	20
Andrews, William T.	33	Otis, Jr., George	20
Parkman, Francis	33	Ripley, George	20
Cooke, Jr., Josiah P.	31		

Table 2.7: Readers with Twenty or More Charges in Experimental Philosophy

Subject	Top Reader	Charges
Astronomy	Miles, Solomon	17
Botany	Boott, John W.	26
Chemistry	Dexter, Franklin	48
Conchology	Tuckerman, Edward	17
Entomology	Parsons, Theophilus	13
Geology	Martin, Enoch	18
Hydraulics	Parris, Alexander	21
Mathematics	Sherwin, Thomas	26
Mechanics	Parris, Alexander	42
Mineralogy	Parkman, Francis	8
Ornithology	Goodrich, Samuel	9
Patents	Eddy, Robert H.	23
Pyronomics	Wyman, Rufus	12
Zoology	Andrews, William	27

Table 2.8: Top Reader in Each Subject Matter Category

The lists of the top five readers and top five titles in each of the subject matter categories can be found in Appendix D.

2.3   Quantifying Reader Interest

In the chapters that follow we will find it useful to identify periods of elevated interest in various topics. In order to accomplish this we will use the rate at which titles relevant to the topic were borrowed as the measure of interest in the topic.

For example, if on average during a particular year books on chemistry were borrowed at the rate of one per day, while books on conchology were borrowed at the rate of one every three days, then we would say that during the year there was a greater interest in chemistry than there was in conchology.

Similarly, if during another year, books on geology were borrowed at the rate of a book every other day during the first six months of the year, but at the rate of one book per week during the last six months, then we would say that there was a greater interest in geology at the beginning of the year than there was at the end of the year.

To illustrate the computation of this measure of interest, consider the overall interest in experimental philosophy by readers at the Boston Athenæum from 1827 to 1850. Figure 2.6 is a cumulative plot of the 3,741[15] charges of titles in experimental philosophy between between January 1, 1827, and December 31, 1850.

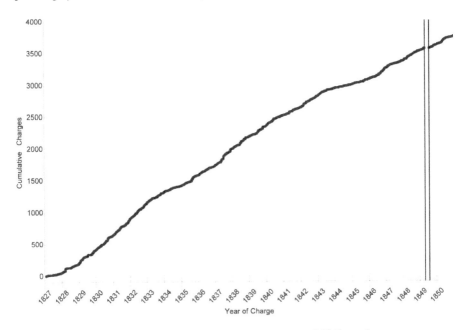

Figure 2.6: Plot of Charges in Experimental Philosophy

To refresh the reader's familiarity with cumulative plots, Figure 2.6 can be read as follows:

1. Select a date on the horizontal **Year of Charge** axis.

2. Go straight up to the cumulative charges curve.

3. Go straight to the left to the vertical **Cumulative Charges** axis.

4. The number read off of the **Cumulative Charges** axis is the total number of charges between January 1, 1827, and the date you selected on the **Year of Charge** axis.

The rate at which books are being borrowed is thus captured by the steepness of the cumulative charges curve. When the curve is very steep, a small change on the horizontal time axis is reflected in a large change on the vertical charge axis indicating

---

[15]The twenty charges for 1851 are excluded.

that a large number of charges were being made per unit of time. Alternatively, when the cumulative charges curve is nearly flat a small change on horizontal time axis maps to a very small change on the vertical charge axis meaning that books are being charged at a low rate per unit of time.

As an example of an anecdotal fact caught in the cumulative charges plot, the gap in borrowing at the far right of Figure 2.6 marks the hiatus in book borrowing from the end of April, 1849, to the end of July, 1849, when the entire Athenæum collection was being moved from 22 Pearl Street to its new home at 10½ Beacon Street.

2.3.1 Computing the Level of Interest

There were 3,741 charges of titles in experimental philosophy in the 8,765 days between January 1, 1827, and December 31, 1850. This works out to about a charge every other day or, what is the same thing, a rate of ½ a charge per day. Using this charging rate, we would estimate that $100/2 = 50$ charges were made in the first hundred days of 1827 and that $200/2 = 100$ charges were made in the first two hundred days of 1827.

If we let $C(d)$ denote the number of charges between January 1, 1827, and d days after January 1, 1827, then these two estimates could be expressed as $C(100) = 50$, $C(200) = 100$, and in general $C(d)$ could be expressed as

$$C(d) = \frac{1}{2} \times d \qquad\qquad (2.1)$$

The multiplier of d in the definition of $C(d)$ is the daily charge rate, one-half a charge per day. It is also our measure of the level of interest.

Using equation (2.1) we can estimate the number of charges of titles in experimental philosophy during the first five years of the study, that is over the 1,826 days from January 1, 1827, to December 31, 1831. To wit,

$$\frac{1}{2} \times 1826 = 913.$$

The actual figure is 934 so our estimate isn't that bad.

Figure 2.7 superimposes this quantification of the overall interest in experimental philosophy on the actual data.

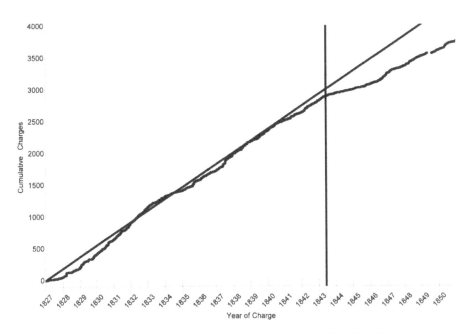

Figure 2.7: Quantifying Interest in Experimental Philosophy

Except perhaps at the very beginning, this measure of interest does pretty well up to around June of 1843 at which time borrowing falls off and our estimate is noticeably too large. Eye-balling the plot from June 1843 onward, the borrowing of books in experimental philosophy seems to have fallen off to a rate of about one book every three days as compared to a rate a book every two days up to that time. This just says what has already noted, namely that reading in experimental philosophy fell off in the mid 1840s. But our estimate has put a number on the steepness of the cumulative charges curve over a period of time and given us a quantitative measure of level of interest during that period—in the case at hand, ½ a book a day.

2.3.2   Periods of Elevated Interest

Table 2.9 is an instance of the tabular format that will be used in the following chapters to call attention to periods of elevated interest in a topic.

Period	Start	End	Readers	Charges	Interest
	Jan. 1827	Dec. 1850	388	3,742	427
A	Jan. 1827	June 1843	280	2,887	500

Table 2.9: Periods of Elevated Interest Experimental Philosophy

The first row of these tables contains the counts of total readers and charges in the topic of the table from the start to the end of the study. Each of the following rows in the table contain the start and end date of a period of elevated interest, the count of the readers and charges during the period.

The value in the right-most **Interest** column is, for easy reading purposes, one thousand times the rate of charging. For the top row this is just the total number of charges, 427 in this case, divided by the number of days between January 1, 1827, and December 31, 1850, namely 8,766. For the remaining rows the value is one thousand times the slope of the line fit to the cumulative charge counts over the row's period and thus can be interpreted one thousand times the rate at which books were being borrowed during the period. Here, $1000 \times \frac{1}{2} = 500$. Or, what is nearly the same thing, one thousand times the number of charges in the **Charges** column by the number of days between the **Start** date and the **End** date.

For titles in the study, there is only one period of elevated interest so value in the **Interest** column of Table 2.9 is not of much, well, interest. When there are two or more periods of elevated interest, the values in the **Interest** column can be used to compare them; a period with a greater interest value being a period with a greater interest in the topic than a period with a lower value. See, for example, the periods of elevated interest table for chemistry immediately below.

The letter in the first column of these tables is used to refer to a specific period of elevated interest in the running text. For example, a discussion of readers in experimental philosophy based on Table 2.9 might include the sentence "The period of highest interest in experimental philosophy was Period **A**."

2.3.3 Comparing Levels of Interest

The values in the **Interest** column quantify the rate of charging so if an interest value for one period is greater than the interest value for another period this is taken to mean that the intensity of interest in the topic of the table during the first period was greater than the intensity of interest during the second period. This is the typical use of these periods of elevated interest tables, that is comparing interest levels between various periods within a single table.

When interest values in two different tables are compared, some interpretative caution may be in order. It remains true, of course, that when any interest value is greater than any another it always means that charges per unit of time of titles of the first was greater than charges per unit of time of the second.

But since the time over which charges are made is the same for all of these tables—the 8,766 days between January 1, 1827, and December 31, 1850—the rates depend on the total number of charges made during this period. This total is the same for interest values within a single table but different for interest values in different tables.

For example, the charging rate for titles in, say, chemistry will always be less than the charging rate for titles in experimental philosophy since chemistry charges are included in experimental philosophy charges. If one wonders how interest in chemistry during some period relative to overall interest in chemistry compares to interest in experimental philosophy relative to overall interest in experimental philosophy then one need only divide the values in the **Interest** column by the total number of volumes charged in the first row of the respective tables.

For example, Table 2.10 shows the two periods of elevated interest in chemistry.

Period	Start	End	Readers	Charges	Interest
	Jan. 1827	Dec. 1850	130	459	52
A	Nov. 1831	Dec. 1832	19	59	150
B	Feb. 1842	May 1843	24	57	126

Table 2.10: Periods of Elevated Interest in Chemistry

To compare the relative interest in chemistry during chemistry's period **A** to overall interest in chemistry to relative interest in experimental philosophy during same period one would compare

$$\frac{150}{459} \text{ versus } \frac{500}{3742},$$

that is 0.33 versus 0.13, and conclude that during chemistry's period **A** there was more interest in chemistry than in experimental philosophy in general. The same method can be use to compare relative interest values in any two such tables.

This is a rather arcane point and is only mentioned as head-up when interest levels are being compared between elevated interest tables. There are two possible comparisons and thus two possible interpretations.

2.4   Summary

While it may not thought of as such today, in the first half of the nineteenth century the Boston Athenæum was a scientific and technical library just as much as it was a library of the arts and letters. The membership of the institution included leaders of the engineering community, luminaries in the medical profession, managers of massive construction projects, as well as renowned chemists, botanists, mathematicians, architects, and educators. Athenæum members were active contributors to the scientific and technical fields in which they read although, as will be seen, their contributions were not always expressed in writing.

In these discourses, where our design is only to inform readers, not to delight or persuade them, perspicuity ought to be esteemed at least one of the best qualifications of a style.

A Proëmical Essay
Robert Boyle

Chapter 3

Hydraulics and Steam Engines

3.1 Dr. John C. Warren and Water for Boston

On November 23, 1825, one of our readers, Dr. John C. Warren, wrote to the mayor of Boston, Josiah Quincy, as follows:

> In the course of their practice, [Boston physicians] say they have noticed many diseases to be relieved and cured by an exchange of the common spring water for soft water of the aqueduct, or distilled water. Hence they have been led to the opinion, that many complaints of an obscure origin, owe their existence to the qualities of the common spring water of Boston. The introduction of an ample supply of pure water, would therefore, they apprehend, contribute much to the health of the place, and prove one of the greatest blessings which could be bestowed on this city. [42, p. 6]

Warren was head of surgery at the hospital he had helped found, the venerable MGH, Massachusetts General.

At the time, water for domestic use in Boston came from a hodge-podge of cisterns and backyard artesian wells in addition to leaky wooden pipe networks such as those of the Old Conduit, the Jamaica Pond Aqueduct, and the Boston Hydraulic Corporation. Many locals were satisfied with the current situation but it was becoming clear to the town fathers that a patchwork water supply strategy was not going to support the urban growth they were witnessing and were, in fact, encouraging.

In addition to the health concerns voiced by Dr. Warren, there was a second, more urgent need for a reliable and copious water supply: putting out fires. Worry about an urban conflagration was nothing new. In the Preface to Evald Rink's checklist of early American technical publications, historian of technology, Eugene S. Ferguson, describes the earliest entry as as follows:

[A] Massachusetts broadside of 1683, making public a law to encourage
the building of brick and stone houses, instead of wood, in the city of
Boston. This, it may be noted, was less than twenty years after the great
London fire of 1666. [202, p. viii]

In the first quarter of the nineteenth century, out-of-control fires in Boston were
becoming alarmingly frequent: the Charles Street fire on July 7, 1824, the Kirby
Street Fire on April 17, 1825, and the Court Street Fire on November 10, 1825, to
name just a few. [43] The great London fire of 1666 may have been forgotten by then
but Boston's great Exchange House fire of November 2, 1818, was certainly fresh
in everybody's memory. The haunting fear of fire had convinced Mayor Quincy to
organize the Boston fire department on April 29, 1826, but there is no particular
advantage to having a fire department if it doesn't have the water to put out fires.

3.2   Purposeful Reading about Hydraulics

Introducing pure water may have been the most contentious Boston forum in which
hydraulics were considered by our readers in experimental philosophy but it was by
no means the only one. The movement of water over and under water wheels and
through canal locks was an active topic of experimentation and discussion. That said,
it is not unlikely that the two periods of elevated interest in hydraulics—the first from
December 1835, to June 1838, and the second in February and March of 1843—can
both be attributed to periods of elevated intensity in the debate over finding a supply
of pure water for Boston. Figure 3.1 is the plot of cumulative charges in hydraulics
and Table 3.1 tabulates these two periods of elevated interest.

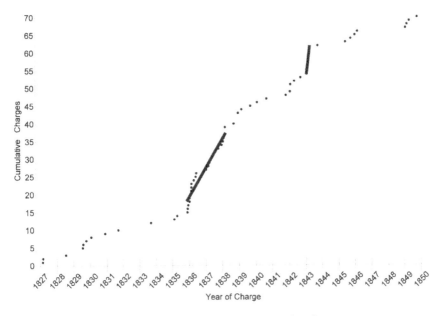

Figure 3.1: Plot of Charges in Hydraulics

Period	Start	End	Readers	Charges	Interest
	Jan. 1827	Dec. 1850	64	161	18
A	Dec. 1835	June 1838	11	25	22
B	Feb. 1843	Mar. 1843	8	8	111

Table 3.1: Periods of Elevated Interest in Hydraulics

The titles of the seventeen titles in hydraulics found in the books borrowed registers along with the number of readers and charges for each are given in Table 3.2. The names of readers who borrowed two or more of these titles are listed in Table 3.3.

Author	Short Title	Readers	Charges
Matthews	*Hydraulia*	12	19
Brisson	*Ponts et Chaussées*[1]	2	11
Ewbank	*Account of Hydraulic and Other Machines*	10	10
DuBuat	*Principes d'Hydraulique*	5	6
Prony	*Recherches Physico-Mathématiques*	5	6
Clare	*Treatise on the Motion of Fluids*	2	3
Aubuisson	*Traité Hydralique*	2	3
Bland	*Hydrostaticks*	2	2
Prony	*Nouvelle Architecture Hydraulique*	2	2
Johnstone	*Draining Land*	2	2
Beaufoy	*Nautical and Hydraulic Experiments*	1	1
Edwards	*Mechanics of Fluids, and Hydraulic Architecture*	1	1
Bernoulli	*Hydrodynamica*	1	1
Cessart	*Description des Travaux Hydrauliques*	1	1
Vince	*The Principles of Hydrostatics*	1	1
Bland	*Elements of Hydrostatics*	1	1
Darwin	*Draining Morasses*	1	1

Table 3.2: Charges and Readers of Books on Hydraulics

---

[1] This is a two volume work. The full title of the first volume is *Recueil de 245 Dessins ou Feuilles de textes relatifs à l'Art de l'Ingénieur, extraits de la Première Collection terminée en 1820, et Lithographiés à l'École Royale des Ponts et Chaussées* and the title of the second volume is *Recueil de 239 Dessins ou Feuilles de textes relatifs à l'Art de l'Ingénieur, extraits de la Première Collection terminée en 1825, et Lithographiés à l'École Royale des Ponts et Chaussées*. The two volumes go by the short title *Ponts et Chaussées* in the Athenæum's catalog. The book is discussed in Chapter 7.

Reader	Charges	Reader	Charges
Parris, Alexander	12	Colburn, Warren	3
Eddy, Robert H.	9	Jackson, Patrick Tracy	2
Boott, John W.	4	Grigg, William	2
Gray, Horace	4	Boyden, Uriah A.	2
Heard, John, Jr.	3	Warren, Edward	2

Table 3.3: Top Readers in Hydraulics

3.2.1 Spot Pond to Bunker Hill

You might have thought that providing potable water to the citizens of antebellum Boston would garner the instant and unanimous support of the citizenry. If you did, you don't know Boston. Nathaniel Bradlee tells the story of the half-century scrum of competing interests in his *History of the Introduction of Pure Water into the City of Boston, with a Description of its Cochituate Water Works.* [42] The names of ten individuals listed in the Athenæum's books borrowed registers appear in Bradlee's history: Cyrus Alger, Edward Everett, Dr. John C. Warren, Charles T. Jackson, Robert H. Eddy, Daniel Treadwell, Nathan Hale, Patrick Tracy Jackson, Lemuel Shattuck, and Edward Brooks.

Cyrus Alger was the first of our readers to grapple with the problem of supplying Boston with fresh water. In 1816 he explored the possibility of sourcing the water from Spot Pond in nearby Stoneham but decided that it wasn't expedient for reasons unknown. [124, p. 123] There the matter stood until May 19, 1825, when a committee was chosen by the city council to see what could be done. Another of our readers, Daniel Treadwell, was appointed by the committee "to ascertain the practicability of supplying the city with good water for the domestic use of the inhabitants, as well as for extinguishing fires, and for all the general purposes of comfort and cleanliness." [42, p. 4] It wasn't preordained that the water would be supplied by a governmental entity but it was clear to everyone that a single citywide system was what was called for so consolidation of the existing do-it-yourself supplies was going to be necessary.

It would be two decades of dueling experts and one water committee after another, until Water Day, October 25, 1848, when "the gate [to the Fountain on the Common] was gradually opened, and the water began to rise in a strong column, six inches diameter, increasing rapidly in height, until it reached an elevation of eighty feet." [2]

The hot button issue, the one that created the most turmoil in the time between Gould's letter to Mayor Quincy and Water Day, was the source of the water. There was

---

[2]See [42, p. 78].

a solid consensus that gravity was the best—that is to say, cheapest and most reliable—means of moving the water to downtown Boston from whatever source was chosen. But this lone point of agreement didn't serve to reduce the short list of candidates since most of the city of Boston is effectively at sea level.

Daniel Treadwell delivered his report to the first water committee on November 4, 1825. Backing up Alger's field work, Treadwell argued that it would in fact be expedient to tap into Spot Pond:

> The water of Charles River is at all times abundant for the supply of the city, although it is not sufficiently elevated to be distributed, without being at first raised by artificial means. But Spot Pond is 140 feet above the tidewater, and consequently its water may be brought to the highest land in the city, by an aqueduct, without any further elevation. [233, p. 5]

Treadwell had visited the steam pumping system—his 'artificial means'—in Philadelphia and had been impressed by what he saw. The Philadelphia steam engine blew up shortly after Treadwell filed his report and this may have cast a pall on his Spot Pond recommendation.

Boston's engineering *éminence grise*, Loammi Baldwin, Jr.,[3] was next to try to resolve the water source issue. In a report delivered to the the next water committee a decade later Baldwin wrote:

> From all the sources I have examined in the vicinity of Boston, as before stated, the most eligible are those of Farm and Shakum Ponds in Framingham, together with incidental ones dependent upon them; and Long Pond in Natick. [9, p. 45]

Baldwin's recommendation included an addenda by another of our readers, Charles T. Jackson, that added Jamaica Pond to Baldwin's list as it could, Jackson opined, "distribute ten times the quantity of water that had hitherto been used."

3.2.1.1   The Report of Robert H. Eddy

Next up was Robert H. Eddy. Eddy was an energetic and prosperous Boston civil engineer, educated at Boston English and tutored in civil engineering by the aforementioned Loammi Baldwin. He contributed to numerous construction projects in and around Boston including the Middlesex and Merrimack River Canals and the dry dock at the Charlestown Navy Yard. As the Civil Engineer to the East Boston Company, he had surveyed that section of Boston, producing maps of the area, and overseeing the construction of wharves, bridges, and roads. [78, p. 206] On January 14, 1836, Eddy was charged by yet another water committee to...

---

[3] Henceforth just 'Loammi Baldwin'.

...survey Horn and other Ponds emptying into Mystic Pond; also Spy and Fresh Ponds in Cambridge; and on the 21st of April, requested him to report on the cost of the introduction of the waters of Spot and Mystic Ponds into the city. [77, p. 3]

A little more than a month after accepting the water committee's charge, Eddy began to borrow books on urban water supplies from the Boston Athenæum. Most notable among these was *Hydraulia; An Historical and Descriptive Account of the Water Works of London and the Contrivances for Supplying Other Great Cities, in Different Ages and Countries* by William Matthews. *Hydraulia* [171] is a 454-page tome covering all aspects—historic, social, engineering, scientific, etc.—of supplying water to urban concentrations using London as a case study. One can hardly imagine a book more suited to Eddy's charter than this one. In fact, *Hydraulia* seems to have been the preferred reference on the topic as it was borrowed by fourteen other of our readers. Two years after he finished his report, Eddy borrowed the book twice again; on February 24, 1838, and on October 4, 1843.

Eddy's deliverable, *Report on the Introduction of Soft Water into the City of Boston*, was turned into the water committee promptly on June 13, 1836. It is short, only 26 pages. The published tract [77] containing the report includes three supporting addenda:

- a chemical analysis of water samples from candidate sources performed and reported by Charles T. Jackson[4]

- a letter from George Odiorne on discussions with the owners of Spot Pond

- the articles of incorporation of the Boston Hydraulic Corporation, the principles which were William Sullivan, another of our readers, Daniel P. Parker, and Eddy's father, Caleb Eddy.[5]

In his report, Eddy introduces two terms describing water distribution in Boston that will be found throughout all subsequent discussions of water for Boston and terms that are still in use today: the *high service area* and the *low service area*. Eddy defined the high service area as the 20% of the land area of Boston that is twenty feet or more

---

[4]At the conclusion of his analysis, Jackson muses about another benefit of a clean water supply for the city: "It appears to me that the 'care worn expression' which strangers remark in the countenances of Bostonians is to be imputed, in a great measure, to the disorders produced by the action of impure water in their constitutions." Bradlee doesn't report any improvement in Bostonian demeanors following Water Day.

[5]Two other reports regarding water for Boston are in the collection: *Statement of evidence before the Committee of the Legislature, at the session of 1839, on the petition of the city of Boston, for the introduction of pure soft water* and *Remarks on the present project of the city government for supplying the inhabitants of Boston with pure soft water* by Henry Bromfield Rogers. Rogers was a reader at the Athenæum but not in experimental philosophy.

above high tide. The low service is the remaining 80% of Boston, viz. land that is twenty feet or less above high tide. As Eddy saw it:

> [T]he high service of the city will always be occupied by private dwellings, and the low service by tradesmen, mechanics, artizans [sic], and a portion of the community devoted to manufacturing pursuits. [77, p. 10]

It is not at all obvious, at least to me, what Eddy is imputing about differential use of water. One supposes that, whatever it was, it was obvious to his readers.

In the event, Eddy once again recommended using Spot Pond and, when the need arose, Mystic Pond. Noting that the surface of Spot Pond is forty feet higher than the top of Bunker Hill, he envisioned a reservoir atop Bunker Hill which would be fed by gravity from Spot Pond and which, in turn, would feed the high service area by gravity. Figure 3.2 is an 1851 map of part of the high service area which shows both the Beacon Hill (not Bunker Hill) reservoir behind the state house and the Boston Athenæum on Beacon Street.

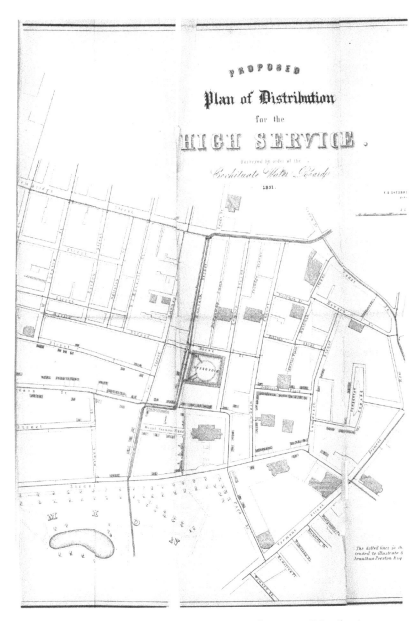

Figure 3.2: Boston's High Service Area for Water Distribution

3.2.1.2   Sizing the Pipe from Spot Pond to Bunker Hill

Like his mentor, Loammi Baldwin, before him, Eddy was obliged to cost out his recommendation and, in particular, compute how much pipe he'd need to carry four million gallons of water per day from Spot Pond to his proposed Bunker Hill reservoir. Spot Pond is 6½ miles from Bunker Hill so he knows how many linear feet of pipe he'll need. To work up a cost for the pipe he only needs to know how big the pipe would have to be.

Water pipe is measured in how many cubic feet of water it can carry per second so Eddy starts out by converting his gallons per day recommendation to cubic feet per second. There are 7.48 gallons of water in a cubic foot of water:

$$7.48 \frac{\text{gallons}}{\text{cubic foot}}.$$

To get the number of cubic feet of water per gallon, we just flip this over:

$$\frac{1}{7.48} \frac{\text{cubic feet}}{\text{gallon}}.$$

Now, to convert four million gallons of water per day to cubic feet of water per day we just multiply:

$$\frac{1}{7.48} \frac{\text{cubic feet}}{\text{gallon}} \times 4,000,000 \frac{\text{gallon}}{\text{day}} = \frac{4,000,000}{7.48} \approx 536,000 \frac{\text{cubic feet}}{\text{day}}. \qquad \textbf{(3.1)}$$

An easy way to keep track of units in computations such as these is to treat them as if they were fractions and cancel:

$$\frac{\text{cubic feet}}{\cancel{\text{gallon}}} \times \frac{\cancel{\text{gallon}}}{\text{day}} = \frac{\text{cubic feet}}{\text{day}}.$$

We want cubic feet per second, not cubic feet per day so we have to multiply this last term by the number of days in a second. The unit 'day' will cancel out and we'll have the units we want, cubic feet per second.

There are 86,400 seconds in a day:

$$60 \frac{\text{seconds}}{\cancel{\text{minute}}} \times 60 \frac{\cancel{\text{minutes}}}{\cancel{\text{hour}}} \times 24 \frac{\cancel{\text{hours}}}{\text{day}} = 86,400 \frac{\text{seconds}}{\text{day}}.$$

So, if there are 86,400 seconds in a day, then we can just flip this number over to get the number of days in a second:

$$\frac{1}{86,400} \frac{\text{days}}{\text{second}}.$$

Finally, to get the number of cubic feet of water per second, Eddy multiplies the number of cubic feet per day by the number of days per second to arrive at:

$$536,000 \frac{\text{cubic feet}}{\text{day}} \times \frac{1}{86400} \frac{\text{days}}{\text{second}} \approx 6.17 \frac{\text{cubic feet}}{\text{second}}. \tag{3.2}$$

Now that he has converted the amount of water he wants to the units that pipe manufacturers and hydraulic engineers use to size pipe, Eddy can tap into the hydraulic engineering literature to estimate the size of the pipe he needs. In his report, Eddy describes this next step in his analysis as follows:[6]

> By De Prony's formula, the proper size of the pipe is determined as follows.[7]
>
> $$Q = 38,116\sqrt{D^5 j}$$
>
> Q = discharge per second=6.17 cubic feet.
>
> D = diameter.
>
> $j = \frac{h}{l} = \frac{\text{head}}{\text{length}} = \frac{32}{26400}.$
>
> $h = 32 \quad l = 26400$
>
> From the above formula, we find a pipe 1,849, or say 22 inches in diameter, would be sufficient for our purpose.

Eddy's evaluation of de Prony's formula using the values he gives looks like this:

$$6.17 = 38.116\sqrt{D^5 \frac{32}{26400}}$$

so

$$\frac{6.17}{38.116} = \sqrt{0.0012 D^5}.$$

Squaring both sides and moving 0.0012 to the other side, we have:

$$D^5 = \frac{1}{0.0012} \times \left(\frac{6.17}{38.116}\right)^2 \approx 833 \times (0.000162)^2 \approx 833 \times 0.026 \approx 21.7.$$

Taking the fifth root of both sides we get:

$$D = \sqrt[5]{21.7} \approx 1.849 \text{ feet} \approx 22 \text{ inches}.$$

---

[6] See [77, p. 14].

[7] In this particular passage in his report and in none other, Eddy uses a comma for a decimal point in 38,116 and 1,849. A conjecture as to why he may have done so follows shortly.

The bottom line is that Eddy needs six and one half miles of twenty-two inch pipe to deliver four million gallons of water per day from Spot Pond to downtown Boston.

Eddy borrowed two books by de Prony from the Athenæum—*Recherches Physico-Mathematiques* [195] and *Nouvelle Architecture Hydraulique* [194]—but this was two years after he turned in his report. The formula of de Prony that Eddy cites is however right off page 54 of Charles Storrow's *Treatise on Water-Works for Conveying and Distributing Supplies of Water* [224] published in March of 1835, about a year before Eddy's report.

It is hard to believe that Eddy and Storrow didn't know each other. They were both members of the Boston Athenæum and proteges of Loammi Baldwin and they both worked on hydraulics projects in and around Boston. Like Storrow, Eddy had traveled to France to study hydraulic engineering projects. Writing in the third person in his family genealogy years later, Eddy recalls:

> In 1838 he visited Europe for examination of the public works there, such as docks, canals, bridges, railways, water-works, etc.; was with the late Elie de Beaumont at the meeting of the French Institute, when Arago, the President, announced the discovery of the daguereotype,[*sic*][8]—being the only American present. [78, p. 206]

Storrow studied with de Prony at the École des Ponts et Chaussés. His *Treatise on Water-Works* is by and large a conversion to English units of de Prony's *Recherches Physico-Mathematiques*. The constant 38.116 in de Prony's formula that Eddy cites doesn't appear anywhere in de Prony's work because it is calculated by Storrow from velocities whose units are English not French: feet per second. It's not clear, at least to me, why Eddy didn't cite Storrow. He obviously knew of Storrow's work. Maybe Eddy was just preening a bit when used a comma instead of a decimal point in de Prony's formula. He wanted to make it look as if he had taken the formula out of a French text but the constant gave him away.

So, other than the occasional exploding boiler, what's wrong with Spot Pond? Marcis Kempe, the unofficial historian of the Boston water system, tells the story:

> The problem is topography. Spot Pond is at an elevation 160 feet which is higher than Boston but the distance and low ground to be crossed was extremely limiting. Before about 1820, there were no cast iron pipes and wood couldn't contain high pressure well. The hydraulic need was to serve the whole of the city which includes Beacon Hill at about ninety feet of elevation. So you need to contain over fifty pounds per square inch at sea level if you want to push it there by gravity....[T]he most complete

---

[8] Another reader at the Athenæum, John Whipple, was the proprietor of a daguerreotype portrait studio in Boston and participated in taking the first astronomical photographs at the Harvard Observatory.

Spot Pond proposal showed a gravity aqueduct with modest pressure to a distribution reservoir on Bunker Hill at a ground elevation of only about sixty-five feet. Then pipes would be laid under the Charles and water delivered at about the "trickle" level of the older Jamaica Pond Aqueduct. So Beacon Hill, including the new statehouse, could not be served.[9]

Marcis adds:

When Loammi Baldwin and later engineers got involved after 1830, they steered the planning westward to get water at a higher elevation.

As things turned out, Eddy was wrong on the particulars—it was Long Pond[10] to Beacon Hill not Spot Pond to Bunker Hill—but he was dead right about the architecture of supplying water to Boston—the water would flow by gravity and would be distributed from a downtown reservoir.

3.3   Purposeful Reading about Steam Engines

Judging from the books borrowed registers, there was a growing interest in use of steam in and around Boston during the second quarter of the nineteenth century. Its conversion to a motive force by means of a steam engine is considered here. Its use for heating buildings is considered in the next chapter.

Forty readers generated ninety-three charges against nine books about steam engines. The titles of the nine books along with a count of the number of readers and charges of each are listed in Table 3.5. Figure 3.3 is the cumulative charge plot for these books. The period of slightly elevated interest is from the end of January, 1830, to the end of March, 1837.

---

[9] Personal communication.

[10] Long Pond was renamed Lake Cochituate at the time of the aqueduct's ground breaking in 1846.

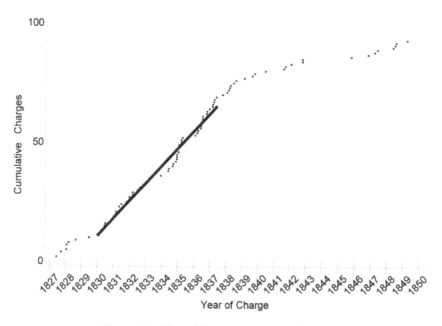

Figure 3.3: Plot of Charges on Steam Engines

Period	Start	End	Readers	Charges	Interest
	Jan. 1827	Dec. 1850	40	93	11
A	Jan. 1830	Mar. 1837	23	59	20

Table 3.4: Periods of Elevated Interest in Steam Engines

Author	Short Title	Readers	Charges
Lardner	*The Steam Engine*	21	24
Tredgold	*Steam Engine Investigation of its Principles*	14	24
Farey	*Treatise on the Steam Engine*	8	23
Stuart	*Descriptive History of the Steam Engine*	9	13
Partington	*Descriptive Account of the Steam Engine*	3	4
Howard	*Report on Steam Carriages*	2	2
Birkbeck	*Steam-Engine*	1	1
Cummings	*Railroads and Steam Carriages*	1	1
Buchanan	*Treatise on Propelling Vessels by Steam*	1	1

Table 3.5: Titles of Books on Steam Engines

3.3.1   Alexander Parris and Lester's Pendulum Steam Engine

Alexander Parris was the top reader of books on steam engines during the single period of elevated interest. His interest during the period can readily be traced to the process of selecting a steam engine for the engine house he was designing for the Charlestown Navy Yard. Parris ended up going with an unusual design, the pendulum steam engine, patented in 1827 by Boston ironmonger Ebenezer Avery Lester .[11] Parris' expression of satisfaction with Lester's engine is included at the end of a long article about the engine in the *Franklin Journal*:

> ALEXANDER PARRIS, Superintendent of the Dry Dock, certifies to the facts before stated, and to the general excellence of the engine; and a similar certificate is given by eight other persons employed there. [163, p. 103]

---

[11] Lester's patent is 4749½X issued April 14, 1827. An X patent is a patent which was issued prior to July 4, 1836, when the Patent Act of 1836 went into effect. This rewrite of the patent law considerably stiffened the requirements to be granted a patent and restarted the patent numbering system. The consequence is that there are two patents numbered 1; 1X dated July 31, 1790, issued to Samuel Hopkins of Philadelphia, Pennsylvania, for an "Improved Method of Making Potash" and US 1 dated July 13, 1836, issued to John Ruggles of Thomaston, Maine, for a "Locomotive Steam-Engine for Rail and Other Roads." The renumbering was not occasioned by the disastrous patent office fire of December 15, 1836, but it may have been a temptation which the fates couldn't resist. In this book the patent numbers and titles rendered in italics such as Josiah Flagg's *1307X* refer to patents that have not been recovered after the 1836 fire. 4749½X is called, for obvious reasons, a fractional patent. There are roughly 100 of these, most but not all of them also X patents. For a detailed history of the early patent office see [73].

Table 3.6 lists titles in experimental philosophy that Parris charged more than once. His most frequent charge was of the *London Mechanics Magazine*. Table 3.7 lists the date and volume number of each of these. Only volumes 4, 16, 19, and 24 were borrowed more than once. The *London Mechanics Magazine* consists primarily of mostly news items on technological developments of the day so it is difficult to say why Parris came back to these particular volumes.

Author	Short Title	Charges
Robertson	*London Mechanics Magazine*	24
Brisson	*Ponts et Chaussées*	8
Farey	*Treatise on the Steam Engine*	5
Tredgold	*Principles of Warming and Ventilating*	4
Parnell	*Treatise on Roads*	2
Matthews	*Hydraulia*	2
Nicholson	*The Operative Mechanic*	2
Robison	*Mechanical Philosophy*	2
Lardner	*The Steam Engine*	2

Table 3.6: Alexander Parris' Charges in Experimental Philosophy

Date	Volume(s)	Date	Volume(s)
Jan. 30, 1837	4	Mar. 9, 1839	28
Feb. 1, 1837	2,4	Jun. 22, 1839	29
Feb. 14, 1837	19, 20, 22	Oct. 5, 1939	30
Mar. 4, 1837	23	Nov. 9, 1839	24
Mar. 24, 1837	18	Feb. 8, 1840	19
Apr. 7, 1837	10	Apr. 11, 1840	31
Jun. 10, 1837	24, 25	Oct. 18, 1842	35
Jul. 27, 1837	16	Dec. 28, 1842	36
Sep. 28, 1837	16, 17	Feb. 18, 1843	34
Oct. 2, 1837	27		

Table 3.7: Alexander Parris' Charges of the *London Mechanics Magazine*

Parris was schooled in architecture by Boston's master builder, Charles Bullfinch, and in engineering by Boston's master engineer, Loammi Baldwin. Around Boston his architectural achievements include the Quincy Market, Cathedral Church of St. Paul, and the Ether Dome in the Bullfinch Building at the Massachusetts General Hospital. But in the technical community, he was an engineer first and an architect second.[12]

Bostonians were not only reading about and installing steam engines, they were building them. Carroll W. Pursell notes that "[b]y 1838 there were eight engine shops in the city" and that "there were forty-four [steam] engine shops in the commonwealth, seventeen of which were in Boston." [196, p. 102]

Pursell goes on to conjecture that the inventor of the novel steam engine that Parris selected for the Charlestown Navy Yard, Ebenezer Lester, was the proprietor of one of the first steam engine shops in town:

> Enginemaking began in Boston in the 1820s: the observant visitor Anne Royall mentioned a steam engine manufactory in the city in 1826 but did not identify it by name. It may have been the establishment of E. A. Lester. [196, p. 102]

Ebenezer Lester was granted fourteen patents between 1805 and 1835 for everything from washing machines and manufacturing sugar to water wheels and nail cutting machines. His oscillating steam engine is, however, likely to be the patent that earned him the greatest return in both money and reputation. Lester was a member of the Boston Athenæum but his name does not appear in the first four volumes of the books borrowed registers. He is listed as a founding subscriber to the Massachusetts Scientific Library Association and was appointed to the Association's subscription committee at the library's organizing meeting in March of 1826. As told in Chapter 2, the Scientific Library was folded into the Athenæum late in 1826. [20, p. 80]

Only the drawing, shown in Figure 3.4,[13] remains of Lester's application for letters patent on his Pendulum Steam Engine. Fortunately, Lester wrote and published a pamphlet describing his invention which was picked up by the technical publications of the day including the *Franklin Journal* [163], Boston's short-lived *Mechanics' Magazine and Journal of Public Internal Improvement* [82], and London's *Mechanics' Magazine.* [83] As the pamphlet was written for a technical audience, it is likely that it included the text of the patent application or words to that effect.

---

[12] Sara Wermiel has written extensively on Parris' engineering projects. See, for example, [249].
[13] ...and on the cover.

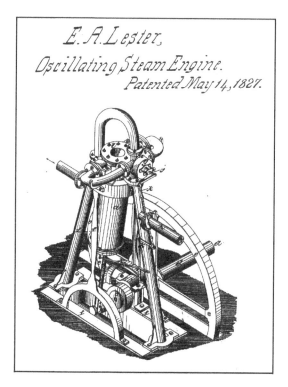

Figure 3.4: Ebenezer Lester's Pendulum Steam Engine

The two-part article on Lester's steam engine that appeared in Boston's *Mechanics'*
Magazine is the longest of the three. The article features an engraving of the engine
made from the patent drawing and what is likely to be a complete reprinting of Lester's
pamphlet since the first paragraph of the article ends with "The pamphlet is as follows:"
Eight certificates testifying to favorable experiences with the engine follow Lester's
promotional description of his engine. There must have been additional testimonials
in the pamphlet since the editors end Lester's text by noting:

> It was deemed the less necessary to add any further certificates, as no fact
> can be stated in relation to any engine of this description, which has ever
> been put into operation, that can, in the slightest degree, detract from,
> or tend to throw the least doubt upon, the results of the operation of the
> engines particularly referred to in these statements. [82, p. 294.]

The first certificate was by Noah Butts, the Chief Engineer at the Charlestown Navy
Yard. The second was by Loammi Baldwin.

Lester was not the first steam engine mechanic to notice that when the up-and-down motion of the piston in a steam engine is converted to the more useful around-and-around motion of a wheel not all of the power generated by the ups-and-downs ends up in the around-and-arounds. Nor was he the first to recapture some of this lost power by freeing the cylinder in which the piston was going up-and-down to rock to-and-fro. Due to the fact that the business end of the engine, the piston, was moving in this manner these engines were called vibrating or oscillating steam engines.[14]

Lester's innovation was to put the axis on which the cylinder rocked at the end of the cylinder rather than in the middle as had its predecessors; hence a pendulum steam engine. Lester claims that the advantage of his pendulum engine over the vibrating and oscillating engines are that 1) the engine can run faster and that 2) gravity helps to power the rocking motion. The first point was an experimentally-proven matter of fact and key a selling point in some steam engine applications. Commentators on Lester's engine were skeptical regarding the second.

There is a coda to the article on Lester's engine in the Boston journal that bears on our considerations. The coda is entitled *Comment on Mr. Lester's Vindication of his Pendulum Steam Engine* [33] and is introduced by the journal's editor as follows:

> The following restrictions upon Mr. Lester's Engine, has been sent us by a scientifick [*sic*] correspondent. As our pages are always open for the free discussion of the merits or demerits of inventions, we freely give it room. [82, p. 294]

The three-page cautionary note is signed with the initials "U.A.B." In fact, there are three other articles in this one-and-only volume of Boston's *Mechanics' Magazine* signed the same way: "On the Effects of blowing Hot and Cold Air into Furnaces" [35], "Letter to the Editor in re 'Combustion of Ashes' " [34], and "Vitality of Matter" [36]. Given the rhetorical tenor of the notes, "U.A.B." is highly likely to be the mark of another of our readers, the quirky Boston mathematician, Uriah A. Boyden. Boyden's critique of Lester's engine had to do with a footnote as well as some fine points of technical terminology in Lester's pamphlet rather than with the engine itself.[15] What is of interest for our purposes are the citations in Boyden's admonishment:

- Patterson's edition of Ferguson's *Lectures on Mechanics*
- Hutton's *Mathematical and Philosophical Dictionary*
- Tredgold's *Treatise on the Steam Engine*
- Nicholson's *Operative Mechanick*
- Dr. Greig's *Encyclopedia*

---

[14]The "Oscillating cylinder steam engine" Wikipedia page explains why power was lost and how oscillating reclaims some of it.

[15]It is not without foundation that Charles Peirce referred to Boyden as "Boston's reproach". [*Writings of Charles S. Peirce*, R 879:30]. Picking fights in public over footnotes is not an endearing trait.

None of these titles are among Boyden's charges at the Athenæum so they were either in his personal library or borrowed from another library in town. These titles are mentioned in passing only to note that citations in published material are another way of gaining insight into what an individual was reading and to what purpose they were doing so. Citations to journal articles in particular clear up any question as to why a journal volume was borrowed and which article was of interest. We will come back to this point in the next chapter.

Commenting on a lengthy footnote in Lester's pamphlet in which Lester bemoans the lack of understanding of the difference between a steam engine and a steam boiler on the part of some writers on steam power, Boyden tuts:

> Dr. Hutton investigated this matter: but Prof. Adrain detected an error
> in his reasoning, and also gave a course of reasoning on the subject; which
> I deem to be likewise erroneous. [33, p. 104]

Robert Adrian was one of American's first theoretical mathematicians. He had edited an American edition of Hutton's *Mathematical and Philosophical Dictionary*. In his comments on Lester's pamphlet in the *Franklin Journal*, the editor, Thomas Jones, acknowledges the errors of the scholars alluded to by Boyden but goes on to write:

> [W]e are not aware that they have prevailed to any extent, nor do we per-
> ceive how it is possible that they should so do excepting in the minds
> of those whose acquaintance with the steam engine is extremely limited.
> [163, p. 99]

Chapter 4

Pyronomics and Hydraulic Turbines

4.1   Rufus Wyman and Heating McLean Hospital

Rufus Wyman was the first superintendent of New England's first hospital. Known originally as Boston's Asylum for the Insane, the name of the hospital was changed to the McLean Asylum for the Insane in 1826 to honor a benefactor, John McLean. When the hospital opened in October of 1818, nine patients and Wyman's family lived together in its only building, a mansion designed and built by Charles Bulfinch for Joseph Barrell.

A new two-story building, euphemistically called the Lodge, was to be added to the asylum in 1826. The Lodge would be home to patients with particularly troublesome behaviours. The apartments on the second story were for "idiots and epileptics, those who were objectionable in their habits or subject to sudden outbreaks of frenzy." The apartments on the first story, called the strong rooms, were for patients "who at times are violent and noisy [and] who defile their rooms in every possible way—the most violent male insane." [260, pp. 21–22].

If you thought that heating and ventilation considerations for an insane asylum would be—for practical purposes—about the same as those for a choir loft, Nina Little in her book about McLean Hospital begs to differ:

> The Lodge was built when Morrill Wyman, Rufus Wyman's second son, was a young man and still making his home at the Asylum, and in later years he remembered some rather interesting details:…'Notwithstanding the ample preparations for ventilation and warming the air of the rooms

and corridors, it was deemed essential that the floor should be well warmed.' To do this effectually and equitably is a matter of no little difficulty. [165, p. 42]

Little goes on to note:

> Central heating of public buildings, still augmented of course by fire-places, was a prime subject of experimentation during the early nine-teenth century and Bullfinch had planned for its installation in both the Asylum wings and the Hospital.

In the event, a 2,000 year old Roman solution was adopted by Wyman: the hypocaust, heated pipes under the floor. Figure 4.1 is the sectional elevation of the Lodge taken from Morrill Wyman's *Early History of McLean's Asylum*. [259]

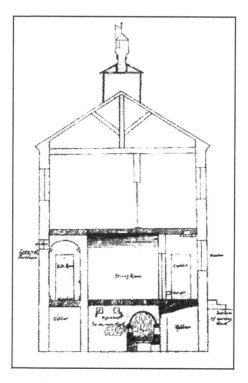

Figure 4.1: Heating the Apartments in McLean's Lodge

As told by both his son and his grandson, Rufus Wyman was deeply involved in designing the Lodge and he paid particular attention to its unique heating and ventilation challenges. Wyman's attention is reflected in his charges of works on heating and ventilation at the Boston Athenæum listed in Table 4.1. There is a gap in Wyman's charges of texts on heating and ventilation from the end of 1839 to the middle of 1842. The charges starting in 1842 more likely reflect books he borrowed on behalf of his son, Morrill, who at the time was working on his own book on heating and ventilation.

Title		Date
Tredgold	*Principles of Warming and Ventilating*	Jan. 31, 1827
Buchanan	*Treatise on the Economy of Fuel*	Feb. 5, 1827
Tredgold	*Principles of Warming and Ventilating*	Feb. 4, 1828
Inman	*Report on Ventilation, Warming*	Nov. 27, 1839
Hood	*Treatise on Warming Buildings*	June 6, 1840
Arnott	*On Warming and Ventilating*	Sept. 10, 1841
Inman	*Report on Ventilation, Warming*	Jan. 25, 1842
Tredgold	*Principles of Warming and Ventilating*	Jan. 25, 1842
Leslie	*Inquiry into the Propagation of Heat*	Feb. 10, 1842
Inman	*Report on Ventilation, Warming*	Apr. 20, 1842
Hood	*Treatise on Warming Buildings*	July 15, 1842

Table 4.1: Rufus Wyman's Charges in Heating and Ventilation

Rufus Wyman was known for his humanist touch in dealing with the mentally ill. He did not countenance bleeding, purgatives, or physical restraints, all of which were practiced at the time. Perhaps this is why there is connection between the Boston Athenæum and the McLean Hospital beyond the borrowing of books on heating and ventilation. Little notes:

Numerous provisions for the enjoyment of the more tranquil or harmless patients, unusual in that day, were also early made: Some of the boarders were quite at liberty to come and go as they pleased. These found their own occupation and amusement; one was a frequent visitor at the reading-room of the Boston Athenaeum, and might have been seen daily among the literary gentlemen who associated there. [260, pp. 23–24]

4.2   Purposeful Reading about Pyronomics

'Pyronomics' is an obsolete term for the science of heating and ventilation. Today we might say 'hydronics' if heating was by means of heated water or spell out the acronym HVAC if we meant heating and ventilation in general but when 'pyronomics' was in vogue there was no AC. Judging from the Google nGram in Figure 4.2, the word came into use shortly before the beginning of this study and died out by the end of the nineteenth century.

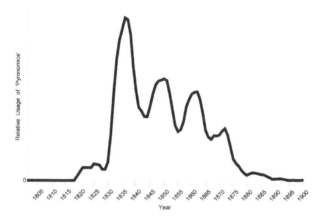

Figure 4.2: Google nGram of 'pyronomics'

Mr. Neil Snodgrass of Johnston, County of Renfrew, Scotland, was voted forty Guineas[1] or a gold medal, his choice, by the Society of Arts for his article, "A Method of Heating Rooms by Steam," in the 1808 volume of *The Repertory of Arts, Manufactures, and Agriculture* [222].[2] In Mr. Snodgrass' third-person telling, it was he who first had the idea of heating rooms by steam. Figure 4.3 is the plate from Snograss' article.

---

[1] Ten ounces of gold or about $38,000 today

[2] Mr. Snodgrass' note also appeared in Volume 27 of Tilloch's *Philosophical Transactions* which was borrowed by Nathaniel Bowditch on March 6, 1830. Bowditch may have been more interested in Mr. Ez. Walker's description of his "new Optical Machine called the Phantasmascope" than in a method for heating rooms. On the other hand, perhaps there was an early March chill in the rooms on 22 Pearl Street.

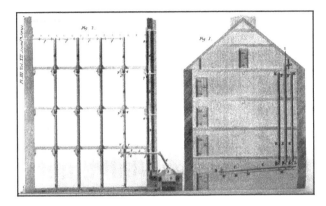

Figure 4.3: Plate from Snodgrass's "A Method of Heating Rooms by Steam"

The use of hearth fires to heat rooms had two widely-recognized shortcomings: much of the heated air went up the chimney rather than into the room and the fire occasionally jumped out of the fireplace and started to use the house and its contents as a source of fuel. A somewhat lesser annoyance was that the radiant heat of the fire over-warmed the sides of the room's occupants facing the fire while providing little comfort to their dark sides. Stoves offered some protection against conflagration but there had to be one in every room to be heated. Steam heating offered the promise of safe (save the occasional exploding boiler) and efficient central heating.

By far the most popular book on steam heating of the day was *Principles of Warming and Ventilating Public Buildings, Dwelling Houses, Manufactories, Hospitals, Hot-houses, Conservatories.* [235] The book's author, Thomas Tredgold, credits Col. William Cook and not Mr. Snodgrass as being the person to first suggest the use of steam for distributing heat inside a building. Col. Cook's short note in the *Philosophical Transactions of the Royal Society* of 1744 [66] consists of the illustration shown in Figure 4.4 with the caption "An Engine for Giving a Sufficient Heat to all the Rooms in a House from the Kitchen Fire."[3]

---

[3]If warming rooms weren't of interest, Cook's note also covers "a method of preventing ships from leaking, whose bottoms are eaten by worms." All of this in a mere three pages.

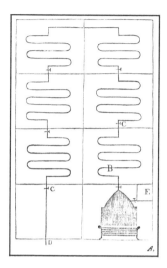

Figure 4.4: Col. Cook's Method of Steam Heating Rooms

After reviewing early patents for steam heating issued to John Hoyle in 1791 and
Joseph Green in 1793 and noting Snodgrass' contribution, Tredgold expresses an opinion about why Cook's pioneering work didn't result in wide adoption:

> Col. Cook's idea was neglected, no doubt, because it promised too much.
> Whoever attempted to warm a large suite of apartments by the spare heat
> of a kitchen first would fail; because so small a quantity of heat is quite in-
> adequate to produce such an effect; but when revived with less pretension,
> steam was found to be a convenient and economical means of distributing
> heat. [235, p. 12]

Sixty-five of our readers generated 130 charges against ten titles on heating and
ventilation. Figure 4.5 is the cumulative charge plot for pyronomics. Table 4.2 lists
the two periods of elevated interest and Table 4.3 lists the ten titles and the number of
readers and charges for each. Rufus Wyman and Horace Gray were the two readers
most interested in the topic with eleven and nine charges respectively.

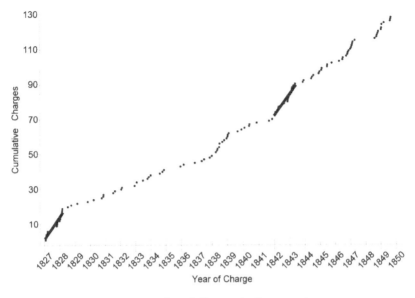

Figure 4.5: Plot of Charges in Pyronomics

Period	Start	End	Readers	Charges	Interest
	Jan. 1827	Dec. 1850	65	130	15
A	Jan. 1827	Feb. 1828	12	19	35
B	Jan. 1842	May 1843	10	20	33

Table 4.2: Periods of Elevated Interest in Pyronomics

Author	Short Title	Readers	Charges
Tredgold	*Principles of Warming and Ventilating*	25	39
Arnott	*On Warming and Ventilating*	14	18
Wyman	*Treatise on Ventilation*	14	16
Leslie	*Experimental Inquiry into the Propagation of Heat*	9	13
Hood	*Treatise on Warming Buildings*	10	10
Buchanan	*Treatise on the Economy of Fuel*	9	10
Reid	*Theory and Practice of Ventilation*	9	9
Inman	*Report on Ventilation, Warming*	5	7
Bell	*Practical Methods of Ventilating Buildings*	7	7
Lardner	*Treatise on Heat*	1	1

Table 4.3: Popular Titles in Pyronomics

It is worth noting in passing that Tredgold wrote on many technical topics and was a popular author among our readers. Forty of them generated seventy-seven charges against the five titles by Tredgold listed in Table 4.4. One needn't be very far out on the limb to conclude that there was something about the way that the Tredgold wrote that appealed to Athenæum readers. The discussion of Tredgold's *Principles of Warming and Ventilating* in the next section suggests what it may have been.

Short Title	Readers	Charges
Principles of Warming and Ventilating	25	39
Steam Engine Investigation of its Principles	14	24
Essay on the Strength of Cast Iron	6	8
Treatise on Rail-Roads and Carriages	3	3
Elementary Principles of Carpentry	3	3

Table 4.4: Readers and Charges of Books by Thomas Tredgold

4.2.1 Tredgold's Experimental Essay for Heating a Room

Chapter III of Thomas Tredgold's *Principles of Warming and Ventilating* is titled "Of the Effect of Steam in distributing Heat, and the Expenditure of Fuel to produce a given Effect."[4] The chapter begins as follows:

> The object of this chapter is chiefly to determine the quantity of steam that will be necessary to heat a given bulk of air; and the proportion of surface of steam vessel or pipe, that will afford any proposed degree of heat in a given time. [235, p. 48]

to which Tredgold editorially adds:

> The usual rules adopted by practical writers are very erroneous.

Chapter III is a quintessential practitioner's experimental essay so it is worthwhile to take a moment to step through it. The question we imagine to be in the reader's mind, the question that caused him to take Tredgold's book off the shelf, might be something like: "How much would it cost to heat my new office with steam?"

Tredgold answers this question for a particular set of parameters—the type of fuel used to make steam, the kind of pipes used to warm the air, the number of windows in the room, what's going on in the room, and so forth. He goes into great detail as he analyzes this specific problem set but he is always careful, as is required for a practitioner's experimental essay, to describe how the details can be adjusted and adapted to fit other situations, in particular and most importantly the reader's new office.

The keystone of Tredgold's essay is his choice of a practical and intuitive numerical value on which to base his analysis. He calls this value the *effect of fuel* and he defines it as follows:

> In order to compare the effects of different kinds of fuel, some convenient measure of effect should be adopted; not only for the purpose of lessening the trouble of calculation, but also to render it more clear and intelligible. I shall, therefore, without regarding the measures of effect employed by others, adopt one of my own, which I have found useful in this and other inquiries of similar nature. I take, as the measure of the effect of a fuel, the quantity, in pounds avoirdupois, which will raise the temperature of a cubic foot of water one degree of Fahrenheit's thermometer. [235, p. 24]

Notice that Tredgold's effect of fuel doesn't mention any specific fuel. It is a property of whatever fuel you'd like to use. Right from the start, if you can figure out how many pounds of, say, old newspapers are needed to raise a cubic foot of water one

---

[4]Chapters I and II cover the history of heating and the general advantages of steam over stoves and open fires while chapters IV and beyond are devoted to special heating and ventilation situations such as churches, cotton mills, hospitals, and greenhouses, as well as some ancillary topics such as using steam for drying and the construction of boilers.

degree Fahrenheit,[5] then you can adapt Tredgold's discourse to heating rooms old newspapers. Tredgold, understanding that the reader from the very start will wish to adapt his running example to their own situation, helpfully includes a table, Table 4.5, that lists the effect of fuel for some popular combustibles.

Kind of Fuel	Effect of Fuel
Newcastle coal	0.0075
Splint coal	0.0075
Staffordshire coal	0.0100
Culm	0.0196
Dry pine	0.0172
Dry beech	0.0242
Dry oak	0.0265
Peat	0.0475
Charcoal	0.0095
Coke	0.0069
Charred Peat	0.0205

Table 4.5: Effect of Fuel for Various Kinds of Fuels

Sadly, old newspapers are not on Tredgold's list but he does cite the experiment conducted by James Watt that established the effect of heat for Newcastle coal:

> Mr. Watt finds that, with the most judiciously constructed furnaces, it requires 8 feet of surface of the boiler to be exposed to the action of the fire and flame to boil off a cubic foot of water in an hour, and that a bushel of Newcastle coals so applied will boil off from 8 to 12 cubic feet. [203, p. 147]

Following this reference, you can replicate Watt's experiment with old newspapers establish their effect of fuel coefficient and continue to follow along with Tredgold's analysis.

After discussing the pros and cons of various fuels Tredgold settles on Newcastle caking coal for his case study. Newcastle coal was abundant, it was sold widely, and its physical and chemical properties had been quantified by John Smeaton and Thomas Thomson as well as Watt.

---

[5]Henceforth degrees are all measured on the Fahrenheit scale.

Tredgold's value for the effect of heat of Newcastle coal, 0.0075, comes directly from Watt's experiment. Tredgold describes Watt's experiment and its outcome as follows:

> The mean weight of a bushel of coals being 84 lbs. and, taking 10 cubic feet as the mean effect of a bushel, it will be found equivalent to heating one cubic foot of water one degree, with 0.0075 lbs. of coal. [235, p. 30]

Tredgold's computation of the fuel effect of Newcastle coal based on Watt's experiment is straightforward. He starts with water at what he calls the mean temperature, $52°$, increases its temperature to $212°$ to bring the water to a boil, and then adds another $967°$ to turn the boiling water into steam.[6] Thus, the total change in the temperature of the water going from $52°$ to steam was $(212-52)+967$ degrees. The amount of water heated was 10 cubic feet and the amount of coal used was 84 pounds. The amount of coal needed for heating one cubic foot one degree—the effect of heat of Newcastle coal—is easily computed as follows:

$$\frac{84}{10 \times (212 - 52 + 967)} \approx 0.0075.$$

The effect of heat is a just a number that translates degrees to pounds. There are no units of measure (like degrees or pounds) associated with it.[7] The number of pounds of Newcastle coal needed to increase the temperature of a cubic foot of water by a specific amount is just 0.0075 times the change of the temperature of the water:

$$\begin{matrix}\text{Pounds} \\ \text{of Coal}\end{matrix} = 0.0075 \times \begin{matrix}\text{Degree} \\ \text{Change of a} \\ \text{Cubic Foot} \\ \text{of Water}\end{matrix} \quad\quad (4.1)$$

All well and good you say, but rooms are filled with cubic feet of air not cubic feet of water.

The specific heat of some material is also a unitless number. A specific heat value quantifies how much easier or harder it is to change the temperature of one material as compared to another. The specific heat of air relative to water is 2,850; that is, air is 2,850 times easier to heat than water. Or said another way, the amount of heat needed to raise a cubic foot of water one degree will raise 2,850 cubic feet of air one degree.

---

[6]Turning boiling water at 212 degrees into steam is equivalent to raising the temperature of the water by an additional 967 degrees. This doesn't mean that steam is 967 degrees hotter than boiling water. It simply means that you have put that much more heat into boiling water to turn it into steam and more importantly for the purpose of heating a room the amount of heat that results when the steam turns back into hot water.

[7]In orthodox standards forums it is said to be of dimension one and be measured in units whose symbol is 1. This may help to explain why there is rarely a line waiting to get into these meetings.

Equation (4.1) told us how much coal was needed to change the temperature of a cubic foot of water. To get how much coal is needed to change the temperature of a cubic foot of air the same amount we just have to divide by 2,850:

$$\frac{\text{Pounds}}{\text{of Coal}} = \frac{0.0075}{2850} \times \frac{\text{Degree Change of a Cubic Foot of Air}}{} \qquad (4.2)$$

For example, the amount of air in a 10×10 room with a 10 foot ceiling is $10 \times 10 \times 10 = 1,000$ cubic feet. To raise the temperature of the air in the room from $50°$ to $80°$ will require

$$\frac{0.0075}{2850} \times \frac{\text{Volume}}{\text{of Air}} \times \frac{\text{Degree}}{\text{Change}} = \frac{0.0075}{2850}(10 \times 10 \times 10)(80 - 50)$$
$$= \frac{0.0075}{2850} \times 1000 \times 30$$
$$\approx 0.08 \ \frac{\text{Pounds}}{\text{of Coal}}$$

In general, to heat a room containing A cubic feet of air from an outside temperature of O degrees to an inside temperature of I degrees will require

$$\frac{\text{Pounds}}{\text{of Coal}} = \frac{0.0075}{2850}A(I - O) \qquad (4.3)$$

pounds of Newcastle coal.

We can now use equation (4.1) to compute how much this amount of heat would change the temperature of a cubic foot of water. We just have to divide equation (4.3) by 0.0075. This cancels out the 0.0075 in the numerator of the fraction so we end up with

$$\frac{\text{Degree Change of a Cubic Foot of Water}}{} = \frac{1}{2850}A(I - O). \qquad (4.4)$$

In words, this equation says the following:

Increasing A cubic feet of air to I degrees from O degrees is the same as increasing a cubic foot of water by $\frac{A(I-O)}{2850}$ degrees.

Put this equation in your back pocket. We'll come back to it in a moment.

The next step in Tredgold's experimental essay is to determine how many feet of pipe filled with steam are needed to maintain the temperature of the room at a comfortable setting of I degrees when the outside temperature is a chilly O degrees. To accomplish this we have to figure out how fast a heated pipe warms the air that comes into contact with it and this depends on what the pipe is made of.

4.2.1.1 Newton's Law of Cooling and the Constant K

You have been served a cup of tea to accompany a dessert that hasn't been served as yet. You like a splash of milk in your tea but you want the tea to be as hot as possible when you get around to sipping it. Do you add the milk immediately or do you wait until dessert is served?

It's a long-eared puzzler. Any tea drinker will tell you to add the milk immediately and will happily explain why. It's simply a matter of Newton's Law of Cooling: the rate at which the tea loses heat depends on how much heat it contains. The hotter the tea the faster it cools. Since adding the cold milk is going to reduce the temperature of the tea by the same amount whether you add it now or when dessert arrives, adding it now means that less heat will be lost while you wait for the Baked Alaska.

Newton's Law of Cooling can be traced to a six-page anonymous note [184] in the 1701 *Philosophical Transactions of the Royal Society*. The research that went into determining that the author of the note was none other than Sir Isaac Newton is a bibliographic detective story of the first order. In a footnote at the end of his own statement of the law of cooling, Tredgold cites Newton as well as other authors on the topic:

> [I]f the temperature of the air to be heated, or if the surface giving off heat, be different at different times, the heat given off in a given portion of time will be directly as the excess of temperature of [the] surface of the body giving off heat. [235, p. 51]

Giving the reader of a practitioner's experimental essay all the information needed to replicate an experiment includes enabling the reader to recreate any physical constants involved. When in comes to heating rooms with hot pipes, one of the physical constants of critical importance is the rate at which the pipe's heated surface warms the air around it. Here's Tredgold's description of the first of three experiments he conducted to determine this constant:

> With a tinned iron cylinder, very slightly tarnished, the surface 79 square inches, and the quantity of water added to the equivalent of specific heat of the cylinder 62.28 cubic inches. Temperature of the room not varying $\frac{1}{4}$ of a degree from $55\frac{1}{2}^{\circ}$ during the trial. It cooled from 181° to 179° in 1.58 seconds; whence we may infer, that the loss of heat is 0.76° per minute when the excess of temperature is $180-55.5 = 124.5^{\circ}$. [235, p. 55]

Note carefully that Tredgold is not computing heat transfer by simply dividing the difference between the start and end temperatures of the pipe by the duration of the experiment. In a footnote he refers to Playfair's heat loss equation,

$$D(1-m)^t = d \qquad (4.5)$$

where D is the start temperature, d is the end temperature, t is the duration, and mD is the heat loss value sought. Tredgold comments that "mD is easily found with the assistance of a table of logarithms."

Tredgold next develops an equation that expresses how many pounds of Newcastle coal you will need to fill a given number of feet of cast iron pipe with steam. Since we already know how to get from degrees to pounds of coal all we need is a way to get from feet to degrees for various kinds of pipe. Here's Tredgold's general setup:

```
Degree
Change of a        Length   ⎛ Degrees    Degrees ⎞
Cubic Foot  = K ×  of Pipe × ⎜ of Pipe  - of Room ⎟        (4.6)
of Water                     ⎝                    ⎠
```

The constant K is the Newtonian secret sauce that captures the heat transfer properties of the particular type of pipe—iron, tin, copper, lead, or whatever—is being used to heat the air in the room.

Referring back to the tinned iron cylinder, here's the formula for K:

```
          Cubic Inches        Heat Loss
            of Water      ×   per Minute
K = ─────────────────────────────────────
            Surface          Excess
      12 ×            ×
             Area            Degrees
```

where

```
Excess     1  ⎛ Starting   Ending  ⎞   Ambient
Degrees  = ─  ⎜ Degrees  + Degrees ⎟ - Degrees
           2  ⎝                    ⎠
```

So for the tinned iron cylinder, Tredgold gets

$$K = \frac{62.28 \times 0.76}{12 \times 79 \times 124.5} = \frac{47.33}{118026} \approx 0.0004$$

Tredgold estimates K for a number of other kinds of pipe and finally settles on 0.000738 for the K value that is right for cast iron pipes. Plug this into equation (4.6) and you get

```
Degree
Change of a              Length   ⎛ Degrees    Degrees ⎞
Cubic Foot  = 0.000738 × of Pipe × ⎜ of Pipe  - of Room ⎟     (4.7)
of Water                           ⎝                    ⎠
```

4.2.1.2 Tredgold's Master Equation

Now, pull equation (**4.2**) out of your back pocket and notice that you have before you two equations for exactly the same numerical quantity:

```
Degree
Change of a
Cubic Foot
of Water
```

Equation (**4.2**) expresses this quantity in terms of the size of the room and the desired change in its temperature. Equation (**4.7**) expresses the exact same quantity in terms of feet of pipe.

Tredgold now does exactly what any reader of a practitioner's experimental essay expects him to do. He sets the two expressions equal to each other.

Putting equation (**4.2**) on the left-hand side and equation (**4.7**) on the right-hand side we get Tredgold's master equation for heating and ventilation:

$$\frac{1}{2850} A(I - O) = KL(P - I) \qquad (\textbf{4.8})$$

where

> A is the number of cubic feet of air you want to heat,
> O is the outside temperature,
> I is the inside temperature,
> K is the heat transfer constant of the pipe you are using,
> L is the number of feet of pipe you are filling with steam, and
> P is the temperature of the surface of the pipe

Take a moment—even if you hated math in high school—to contemplate Tregold's master equation. It is a *tour de force* and worth the effort. Setting aside the constants $1/2850$ and K, on the left-hand side of the equation we have the amount of air in the room, A, and how much we want to change its temperature, $(I - O)$. On the right-hand side of the equation we have the amount of heating pipe in the room, L, and how much hotter the pipe is than the air in the room, $(P - I)$. You can almost see Tredgold's master equation working: heat the pipe on the right-hand side to warm the room on the left-hand side.

This rhetorical framework is a mainstay of many practitioner's experimental essays. The author derives two mathematical expressions for a key quantity early in his discourse and sets them equal to each other. The resulting master equation can be used to express any of the quantities in the equation in terms of all of the others and, in addition, specific instances of the equation can be used to mathematize the discussion of specific problems.

Oliver Evans in his 450-page classic, *The Young Mill-Wright and Miller's Guide*, states his General Law of Mechanical Powers on page 31:

[T]he momentums of the power and weight are always equal when the engine is in equilibrio. [98, p. 31]

An example of an 'engine' is a grist mill. When a grist mill is running smoothly—in equilibrio—the load on the water wheel is equal to the power being generated by the water wheel.

Evans uses his law throughout his text to analyze a wide range of mechanical engineering problems including the causes of friction, the economy of water wheels, and the optimization of gear trains to name a few. He does so by building one equation for the power being supplied to the engine and another equation for the rate at which the engine is performing some task and, citing his law, sets the two equations equal to each other.

Edwin Layton summarizes Evans' method of working from principle to practice as follows:

Evans presented a set of rules for applying scientific methods to technology, which included the discovery of fundamental principles, making deductions from them, and testing these results by experiment. In each case he illustrated these steps with examples drawn from his own career as inventor. In short, Evans' science was practical, hydraulic, and experimental. [161, p. 68]

This explanatory framework could be thought of as a kind of practitioner's law of conservation. Whenever something causes a change in one side of a master equation, then something on the other side of the master equation must also change to keep the two in balance.

Tredgold's master equation illustrates how a practitioner's law differs from a scientist's law. A scientist's law, take Ohm's for example, describes a relationship between physical measurements at the conceptual level of the measurements:

The current through a conductor between two points is directly proportional to the voltage across the two points.

A practitioner's law describes a relationship too but it is a relationship at a practical, sensible level between something you want to achieve—a warm room—and what you can do to achieve it—heat a pipe.

This is not to say that a practitioner's law is any less general than a scientist's law. A practitioner's law simply has a different units of measurement and is motivated by a different set of objectives. Tredgold's law is as independent of the source of the heat and nature of the pipe that carries it in the same way that Ohm's law is independent of the source of the voltage and the nature of the conductor that carries it.

Getting back to warming the room, we've brought it up to the desired temperature but what about ventilation? If people are going to be working or playing in the room, then we need to push the warm, used air out and push cold, fresh air in otherwise...well, you get the picture. Furthermore, how much air we need to push out and let in per hour will certainly depend on how many people are in the room and what they are doing. A accountant tending to his sums is going to require less ventilation than a high-impact aerobics class.

Ventilating the room is the A in Tredgold's master equation. This variable lets us go from heating a given amount of air to heating a flow of air; that is from heating cubic feet of air to heating cubic feet air per minute.

Rearranging equation (4.8) to express the length of the pipe, L, in terms of all of the other variables, Tredgold gets

$$L = \frac{A(I - O)}{2850K(200 - I)} = \frac{A(I - O)}{2.1(200 - I)} \qquad (4.9)$$

where he has taken the surface temperature of the cast iron pipe, P, to be $200°$ and the heat transfer constant, K, to be 0.000738. Equation (4.9) simply says that if you know:

- the temperature outside, O
- the temperature you want inside, I, and
- how many cubic feet of air per minute needed to ventilate the room, A

then you will need to install L feet of cast iron pipe.

At the end of Chapter III Tredgold expresses equation (4.9) in the context of his running example as the following Rule:

> Multiply the cubic feet, per minute, of air to be heated, to supply the ventilation and loss of heat, by the difference between the temperature the room is to be kept at, and that of the external air, in degrees of Fahrenheit's thermometer, and divide the product by 2.1 times the difference between 200 and the temperature of the room: this quotient will give the quantity of surface[8] of cast iron steam-pipe that will be sufficient to maintain the room at the required temperature. [235, p. 59]

---

[8]In the running text Tredgold refers to the unit of L as "feet of surface of steam-pipe." The nominal unit of L is square inches. The conversion between the two depends on the circumference of the pipe being used. If the outside circumference of the pipe is c then you need $\frac{1}{12}\left(\frac{L}{c} - \frac{c}{2\pi}\right)$ linear feet of the pipe to show L square inches of surface.

Tredgold's Rule is a just a summing up. The true utility of Tredgold's Chapter III experimental essay isn't his Rule at all. The Rule, after all, only pertains to his running example. The true value of his essay is that he has given the reader all that is needed to derive their own version of the Rule, one that is tailored precisely to their fuel supply, their room and whatever goes on in the room. If the reader wants to use old newspapers to make steam, glass steam pipes to warm the air, and use the room for a typing pool, then Tredgold has laid out a step-by-step procedure to determine how much pipe will be needed and what the heating bill will be.

In the rest of his book Tredgold analyzes a variety of special cases—hospitals, churches, theatres, cotton mills, orangeries, and so forth. Consider, for example, the variant of the Rule he derives under the topic "Of the Ventilation and Loss of Heat in Public Buildings and Dwelling Houses." After a discussion of inhalation, exhalation, and various other bodily functions inviting air circulation as well as architectural sources of heat loss such as windows and doors, Tredgold posits the following Rule:

> In public buildings, dwelling houses, &c. the quantity of air in cubic feet to be warmed in one minute should be equivalent to four times the number of people the room is intended to contain, added to eleven times the number of external windows and doors, added to one and a half times the area in feet of the glass exposed to the external air. [235, p. 80–81]

The experiments needed to derive the parameters of this rule—'four', 'eleven', and 'one and a half'—are all described in the text. As with the running example of Chapter III, the reader can derive values that adapt the rule to their own situation using Tredgold's derivations for guidance.

Speaking of local contexts, suppose we wanted to compute how much it will cost to keep three bibliophiles in an nineteenth-century office at $10\frac{1}{2}$ Beacon Street with two outside windows at a warm and cozy $72°$ when it is a Boston winter $10°$ outside? We can use Tredgold's master equation to estimate the amount of coal we will need:

$$8 \times 60 \times \frac{4 \times 3 + 11 \times 2}{2850} \times (72 - 10) \approx 355 \; \text{Pounds of Coal} \; \approx 5 \; \text{Bushels of Coal} \quad (4.10)$$

Multiply this amount by the Boston street price of a bushel of coal, 35 cents, to get, for practical purposes, $1.75 per day.

4.2.2   Morrill Wyman's *Treatise on Ventilation*

Positing a reasonable conjecture about which article in a borrowed periodical volume may have interested the volume's borrower is, as has already been noted, a chancy affair. Usually, but not always. When a reader cites an article in a borrowed volume, then it is a pretty good chance that the article was at least one reason why the volume was taken home.

Morrill Wyman, Rufus Wyman's son, grew up at McLean Hospital; not as a patient, let me hastily add, but because his family lived there. On page 171 of his *Practical Treatise on Ventilation*, Morrill Wyman footnotes the following passage

> Dr. Ure has estimated that the power of the fan, measured by the exhaustion produced, is thirty-eight times that of the ventilating chimney consuming the same amount of fuel. [256, p. 171]

with the following citation:

Mechanics' Magazine, Vol. XXVII, p. 22

Sure enough, on page 515 of Volume II of the books borrowed register allocated to Morril's father, Rufus, we find a charge of volume twenty-seven of the *Mechanics' Magazine* on February 1, 1842:[9]

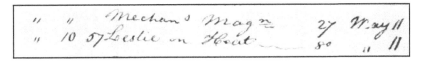

Figure 4.6: Rufus Wyman's Charge of *Mechanics' Magazine*

Turning to the magazine, we find "An Experimental Enquiry into the Modes of Warming and Ventilating Apartments" by Andrew Ure, pp. 21–27 and on page 23 we find:

> Hence, it appears, that the economy of ventilation by the fan is to that by the chimney draught as 66 is to $^{35}/_{20}$, or as 38 to 1.

Now, of course, finding text referenced by a citation where the citation said it would be is how citations are supposed to work so there is no surprise here. But the books borrowed register has contributed a further piece of data that could be of interest to historians: an approximate date on which the published text was being written.

In addition, and in our particular situation, the connection between the footnote in Morrill Wyman's book and the charge to Rufus Wyman's account lends credibility to the conjecture made at the beginning of this chapter that starting in 1839 the father was borrowing books on behalf of his son.

There are a couple of final flourishes regarding Morrill's book and his father's charges. In a number of places in his book, Morrill refers to experiments by Dalton and

---

[9] ...next to a charge of John Leslie's *Experimental Inquiry into the Nature and Propagation of Heat.*

Parent-Duchâtelet but he doesn't say which works of these two authors he consulted. His father charged Dalton's *Chemical Philosophy* on March 2, 1840, and Parent's *Rapport de l'Hygiène Publique* on April 3, 1843, so we can conjecture these were the texts he used. Furthermore, while Morrill doesn't cite the journal in his book, his father borrowed *Annales des Sciences Naturelles* seven times between March 3, 1839, and August 27, 1842, so this journal may also have been another source for Morrill's text.

Finally, as an anecdotal indicator of the Athenæum's role as a forum for the exchange of ideas, on page 259 of his treatise on heating and ventilation Morrill Wyman cites a journal article by another of our readers, George B. Emerson. Emerson's article, *The Schoolmaster. The Proper Character, Studies, and Duties of the Teacher, &c., and the Principles on which Schoolhouses should be Built, Arranged, Warmed, and Ventilated* [94], includes the floor plan for a school for forty-eight pupils. Emerson borrowed Tredgold's book on January 20, 1832, and again on February 8, 1832, and Arnott's *On Warming and Ventilation* [7] on November 30, 1835. It does not strain one's imagination to see them discussing the heating of buildings down on 22 Pearl Street.

4.3   Purposeful Reading about Hydraulic Turbines

Remarking on the art and science of fluid dynamics, Sir Cyril Hinshelwood, the winner of the Nobel Prize in chemistry in 1956, noted:

> Fluid dynamicists were divided into hydraulic engineers who observed what could not be explained, and mathematicians who explained things that could not be observed. [169, p. 122]

The first person to apply differential calculus to an engineering problem was Antoine Parent. [199, p. 207] The problem to which Parent addressed himself was same problem that Athenæum reader Uriah A. Boyden tackled a century and a half later, the efficiency of the water wheel. Parent's analysis, presented to the French Academy of Science in 1704 [186], was a consensus of many of the scholarly mathematicians of the day including d'Alembert, Gravesande, and Colin MacLaurin. [198, p. 276]

It wasn't until experiments were conducted by two experienced wheelwrights, Antoine de Parcieux in France and John Smeaton in England, that it was realized that Parent's formulation was, to be charitable, naïve. The mathematicians had pushed the symbols around with energetic elan for over half a century, solving Parent's equation using all manner of calculus, algebra, and geometry, each and every time coming up with guidance that was at odds with the wheelwrights' experience and counterproductive for practical purposes.

In 1766, Jean Charles Borda published a detailed analysis of the overshot water wheel [21] that brought to light the many shortcomings of Parent's model. Borda's

master equation for the way a water wheel harnessed the power of a moving stream
was the following:

$$
\begin{matrix} \text{Output} \\ \text{of Wheel} \end{matrix} = \begin{matrix} \text{Impact} \\ \text{of Water} \end{matrix} + \begin{matrix} \text{Weight} \\ \text{of Water} \end{matrix} - \begin{matrix} \text{Loss} \\ \text{on Entry} \end{matrix} - \begin{matrix} \text{Loss} \\ \text{on Exit} \end{matrix} . \quad \textbf{(4.11)}
$$

Parent's model of the water wheel only took in account the first of the four terms
in Borda's model; namely, the smashing of the water against the paddles of the water
wheel. Since this force would be about the same whether the water hit a paddle at the
top of the wheel or at the bottom of the wheel, this led Parent and his fellow mathematicians to the wholly erroneous conclusion that an overshot water wheel was just
as efficient as an undershot water wheel. Wheelwrights had known for years that you
could grind a lot more grain with an overshot wheel than you could with an undershot
wheel.

The second term in Borda's model, the weight of the water, explains why the overshot wheel gets more power out of the stream than the undershot wheel. An undershot wheel is driven by just the impact of the water against the paddles—Borda's first
term. An overshot wheel adds to the force of the impact the weight of the water in the
buckets—Borda's second term. It's not as if this is a hidden effect. If all the buckets on
one side of a water wheel are filled with water and all the buckets on the other side are
empty, wouldn't you suspect that full buckets would go down and the empty buckets
would go up?

The third and fourth terms in Borda's model, loss on entry and loss on exit, account
for two less apparent sources of water wheel inefficiency both of which are due to water
turbulence. To maximize the output of a water wheel these two loss terms have to be
driven as close to zero as possible. Translating this into wheel design, it meant that the
water had to enter and leave the wheel without splashing.[10] These two terms, which
at first blush might not seem worth worrying about, led to the water wheel's super-
efficient successor.

Running without splashing is difficult to achieve when the water enters and leaves
the wheel through the same aperture, the mouth of the bucket. In effect, you have
to stop the inward flow once you've filled the bucket and then you have to start an
outward flow when you want to empty the bucket. Once these two sources of power
loss were recognized and modeled by Borda they could seen, measured and reduced.

Poncelet's 1827 experimental water wheel with its curved blades showed the way.
Poncelet's blade arrangement let the water in with little splashing and let it leave with
little velocity relative to the turning wheel. But can we do better? What if the entrance
and the exit were not the same aperture so that the water flowed *through* the wheel

---

[10] Splashing is to hydraulics what noise is to mechanics; noise signals the presence of friction and friction
makes heat not flour.

rather than into and out of it? Inventive tinkering by wheelwrights exploring this possibility independently in France and America, led to a revolution in water wheel design: the hydraulic turbine.[11]

American hydraulic engineers were the leaders in the development of the hydraulic turbine. Norman Smith in his history of the water wheel remarks on this unexpected turn of events:

> The prototype Francis[12] turbines were unquestionably the most scientifically designed, the most thoroughly tested and best constructed up to that time....In a way, it is a rather surprising development. The United States was at this time a new and expanding country, raw and undeveloped in many things, short of skilled craftsmen and with precious little tradition in engineering science. And yet it was here, with unprecedented scientific research and technical precision, that a key class of turbine was evolved. [221, p. 180]

4.3.1   Uriah A. Boyden and the Hydraulic Turbine

Uriah A. Boyden was a Boston-based consulting engineer, mechanical and civil. He received his engineering education on the job, working in the office of Loammi Baldwin, alongside Charles Storrow and readers Alexander Parris, Robert H. Eddy, George M. Dexter. It must have been quite an environment. Figure 4.7 is a picture of Boyden's business card showing the references of some of these colleagues.[13]

---

[11] The history of the water wheel and its displacement by the hydraulic turbine is the canonical case study in the interplay of science and technology. The story is thoroughly told by Terry S. Reynolds in *Stronger than a Hundred Men: A History of the Vertical Water Wheel.* [199] See Danilo Capecchi's article, "Waterwheels in the 18th Century," [52] for a purely mathematical recitation of the history.

[12] James Francis was America's John Smeaton. His practitioner's experimental essay, *Lowell Hydraulic Experiments* [107] was on the book shelf of every American hydraulic engineer in the second half of the nineteenth-century and is still on the bookshelf of many today. See *Waterpower in Lowell: Engineering and Industry in Nineteenth-Century America* [169] for much more about Francis.

[13] ...in particular, Kirk Boott, Loammi Baldwin, Joseph Hayward, Patrick T. Jackson, and J.F. Baldwin.

Figure 4.7: Uriah A. Boyden's Business Card with References[14]

In 1837 Boyden was engaged by James Francis, at the time the chief engineer of the Merrimack Manufacturing Company, to see if more power could somehow be squeezed out of the Merrimack river than was being delivered by the lumbering vertical water wheels. Based on examining Boyden's papers at the Smithsonian, [152] Edwin Layton reports that Boyden had become familiar with the Fourneyron's hydraulic turbine by reading articles by Ellwood Morris in the *Franklin Journal* and that Boyden had examined one of Morris' turbines in operation at the Rockland Cotton Mills in Coatsville, Maryland, in 1844 or 1845. [162, p. 70] Morris' articles are listed in Table 4.6.

Title	Citation
"Remarks on Reaction Water Wheels"	[180]
"On the Friction Dynamometer"	[182]
"Useful Effect of Turbines"	[181]

Table 4.6: Articles on Water Wheels by Edwin Morris

[14]From the Uriah A. Boyden Papers with kind permission of the Archives Center, National Museum of American History, Smithsonian Institution.

The 1843 edition of the *Franklin Journal* also included a three-part article[15] by Arthur Morin, "Experiments on Water-Wheels, having a vertical axis, called Turbines" [179] that couldn't have escaped Boyden's attention.

Boyden's reading in hydraulic turbines as recorded in the books borrowed registers was by no means limited to these articles by Morris and Morin. Appendix F.2 lists Boyden's periodical charges. For each of the periodical charges, save two, the entry in this list includes an article that might have caught Boyden's eye. These selections are pure conjecture, of course, but given Boyden's broad spectrum interest in science and technology as reflected in the Smithsonian's finding guide for his papers, it is hard to go far wrong.

Of particular note in the list of Boyden's periodical readings are the three articles by Leonhard Euler: "Decouverte d'un Nouveau Principe de Mécanique" [95], "Application de la Machine Hydraulique de M. Segner a toutes sortes d'Ouvrages et de ses Avantages sur les autres Machines Hydrauliques dont on se sert Ordinairement" [96], and "Théorie plus complette des Machines qui sont Mises en Mouvement par la Re-action de l'Eau" [97]. These three articles are the well-spring of the two no-splashing terms in Borda's equation and they would also be a point of departure of the career of another renowned hydraulic engineer: Euler's son, Johann. [198, pp. 286–290]

In extolling the utility of calculus in his outgoing Presidential Address to the Indiana Academy of Sciences, Purdue mathematician Clarence Abiathar Waldo chose to denigrate Boyden to bring home his point:[16]

> Fifty years ago an excellent engineer by the name of Uriah Boyden spent weeks in designing the buckets of a water wheel. He obtained correct forms, but by the aid of the calculus a man no more talented naturally may to-day do the same work in two or three hours. [244, p. 48]

Regarding Boyden's mathematical skills, Edwin Layton in a review of the development of the American hydraulic turbine says:

> Boyden and [James] Francis developed a scientific tradition that differed from the European one. Their bias against mathematical theory was not based on ignorance; Boyden, in particular, was an able and creative mathematician, albeit in an archaic, Newtonian geometrical tradition. Boyden, and probably Francis, were familiar with the mathematical theories of the turbine developed by Euler, Poncelet, and Weisbach. [162, p. 76]

---

[15]…translated from the French by Morris.

[16]Waldo's own contribution to the application of mathematics is advising the Indiana legislature on the value of π. Whether or not he used calculus is unrecorded.

Boyden knew calculus perfectly well. He also knew a lot about how water flows. Layton goes on to say:

> But they thought that these theories did not come close enough to reality to be of value to the practical man. [Ibid., pp. 76–77]

Some there are, who desire but to please themselves by the discovery of the causes of the known phænomena; and others would be able to produce new ones, and bring nature to be serviceable to their particular ends, whether of health, or riches, or sensual delight.

A Proëmical Essay
Robert Boyle

Chapter 5

Chemistry and Botany

5.1 Josiah P. Cooke, Jr. and Atomic Weights

In 1836, the Boston polymath, Charles S. Peirce, received the first *summa cum laude* degree in chemistry from Harvard. Of his mentor, Josiah P. Cooke, Jr., Peirce wrote

> The principal precursor of Mendeléef is, as it seems to us, that penetrating intellect, Josiah P. Cooke, who first proved that all the elements were arranged in natural series. [190, p. 169]

Cooke switched from tutoring mathematics to teaching freshman chemistry at Harvard in 1849. The following year, at the age of just twenty-three, Cooke was elected Erving Professor of Chemistry and Mineralogy.[1] In the introduction to his 1854 paper announcing his entry into the race to find what would eventually be called the periodic table of elements, Cooke offers an analogy to explain his decision:

> For a zoölogist to separate the ostrich from the class of birds because it cannot fly, would not be more absurd than it is for a chemist to separate two essentially allied elements, because one has a metallic lustre and the other has not. [67, pp. 237–238]

To Cooke's way of thinking, the organization of the chemical elements and the understanding of how they combine to make compounds had to be solidly grounded in mathematics. Only by way of experimentation and the discovery of numerical relationships among measured values could one gain a useful understanding of qualitative properties. First the quantitative, then the qualitative, not the other way around.

---

[1] The previous holder of the Erving chair had been hung three months before Cook's appointment for the murder of one of our readers, George Parkman.

Fifty years earlier, on October 21, 1803, in an experimental essay read to Manchester Literary and Philosophical Society, the English chemist John Dalton had presciently connected recent results in quantitative chemistry to the ancient and oft derided atomic theory of nature. Dalton's paper, as well as his 1808 book, *A New System of Chemical Philosophy*, initiated research in atomic theory that continues unbroken down to today.[2]

The measurement that sparked Dalton's break-through insight was relative atomic weight: the ratio of the weight of an atom of one element to the weight of one atom of another element. As Dalton sets the stage in *Chemical Philosophy*:

> Now it is one great object of this work, to shew the importance and advantage of ascertaining *the relative weights of the ultimate particles, both of simple and compound bodies, the number of simple elementary particles which constitute one compound particle, and the number of less compound particles which enter into the formation of one more compound particle.*[3] [70, p. 213]

The determination of relative atomic weight is easy on paper but usually quite difficult at the laboratory bench. The experimental protocol is roughly this:

1. Pick a material composed solely of the two elements whose relative atomic weight you wish to determine

2. Split the material into the two elements

3. Weigh the resulting quantity of each

4. Form the ratio of these two weights, taking into account how many atoms of each element comprise the material

5. The resulting number is relative atomic weight of the element whose weight is in the numerator of the ratio to the element whose weight is in the denominator.

This naïve protocol assumes that the chemical formula for the compound that was broken apart—NaCl for salt, for example—was known and this was often not the case.

For example, to determine the relative atomic weight of oxygen to hydrogen you might start with a beaker of H_2O. Decompose the H_2O into its constituent elements, H and O. Now, carefully weigh the volumes of O and the H you created. The atomic weight of O relative to H is then

$$\frac{\text{weight of } O}{\text{weight of } H/2},$$

---

[2]Scientists and engineers were on more congenial terms then than today. One of our readers, Daniel Treadwell, visited Dalton when he, Treadwell, was in England promoting his power press.

[3]Dalton's italics.

where the weight of H is divided by two since there are two atoms of H for every one atom of O in H_2O.[4]

It wasn't long after Dalton's book was published that relative atomic weight values began to appear in the literature and hypotheses bandied about regarding relationships between them. As Cooke coyly notes in the very first sentence of his 1854 paper:

> Numerical relations between the atomic weights of the chemical elements have been very frequently noticed by chemists. [67, p. 387]

Cooke suspected, as did many others, that buried in the atomic weight values and the numerical relationships among them was the eagerly-sought mathematical foundation for a classification system for the chemical elements from which one could deduce their properties and their compounds.

What nobody knew at the time was that atomic weight was mimicking the quantity that three decades later would form a solid basis for classification system for chemical elements: atomic number. But atomic weight was a good enough impersonator that investigations into relations among atomic weights led to predictions of undiscovered elements as well as more efficient chemical manufacturing processes not to mention the discovery of the number behind the mask.

5.2 Purposeful Reading about Chemistry

Recall that the 1807 articles of incorporation of the Boston Athenæum included the following benefit of subscribing to the enterprise:

> [T]he plan of the Athenæum includes a LABORATORY, and an APPARATUS for experiments in chemistry and natural science, for astronomical observations, and geographical improvements, to be used under the director of the corporation. [210, pp. 5–6]

Providing laboratory space and equipment for scientific experimentation was a strong selling point for an Athenæum membership. Educational institutions provided few facilities to acquire hands-on experience with chemistry. The laboratories used by professors as well as by practitioners for their experimentations were likely to have been paid for out of their own pockets and not infrequently installed in their own homes. [251]

---

[4]One of the primary challenges in determining relative atomic weight is finding a material that consists solely of the two elements whose relative atomic weight you want and that you can break apart. If the search doesn't turn up any candidates, determining relative atomic weight becomes a two-stage process: determine the relative atomic of X to Y and the relative atomic weight of Y to Z and then multiply the two together $\frac{X}{Y} \times \frac{Y}{Z}$ to get the relative atomic weight of X to Z. You can readily imagine the amount of measurement uncertainty generated in this two-stage process never mind three- or four-stage processes.

To be sure, scientific courses and public lectures from time to time included experimental demonstrations. In this particular regard, the Athenæum's 1807 articles of incorporation go on to say:

> The LABORATORY and APPARATUS may be used, when it shall be found practicable, for the purpose of lectures on chemistry, natural philosophy, and astronomy. The usefulness of a course of popular instruction upon these and other related subjects, calculated to interest the young of both sexes, and to diffuse as well as extend the knowledge of the laws and operations of nature, need not be displayed. [210, p. 14]

The Boston Athenæum was from its inception just as much natural philosophy society as it was a literary society. This fact further is attested to by its printed catalogs. In the very first catalog compiled by Rev. Joseph McKean in 1810, a mere three years after the Athenæum's founding, roughly 15% of the titles[5] are of scientific topics. The titles of the sixteen chemistry texts in the 1810 catalog are listed in Table 5.1. By the time the 1827 catalog was published there were more than thirty-five titles in just chemistry on the shelf.

Author	Short Title
Bergman	*Physical and Chemical Essays*
Black	*Lectures on the Elements of Chemistry*
Chaptal	*Elements of Chemistry*
Chaptal	*Chimie Appliquée aux Arts*
Ewell	*Discourses on the Elements of Modern Chemistry*
Fourcroy	*General System of Chemical Knowledge*
Fourcroy	*Synoptic Tables*
Henry	*Epitome of Chemistry*
Heron	*Elements of Chemistry*
Hunter	*Dissertatio Chemica de Aetheribus*
Nicholson	*First Principles of Chemistry*
Nicholson	*Dictionary of Chemistry*
Lavoisier	*Méthode de Nomenclature Chimique*
Thomson	*System of Chemistry*
Watson	*Chemical Essays*

Table 5.1: Chemistry Titles in the Athenæum's First Catalog

[5]...setting aside tracts and newspapers.

With 459 charges, chemistry is tied for the most popular subject matter category with geology. The 130 readers in chemistry borrowed fifty-two chemistry titles. Table 5.2 lists the chemistry titles with ten or more charges and Table 5.3 the chemistry readers with more than five charges.

Author	Short Title	Readers	Charges
Liebig	*Organic Chemistry in Agriculture*	35	53
Lavoisier	*Annales de Chimie*	20	52
Faraday	*Chemical Manipulation*	20	37
Turner	*Elements of Chemistry*	23	36
Webster	*Manual of Chemistry*	18	24
Gray	*The Operative Chemist*	21	23
Thenard	*Essay on Chemical Analysis*	10	15
Liebig	*Familiar Letters on Chemistry*	9	12
Parkes	*Chemical Catechism*	9	11
Berzelius	*Analysis of Inorganic Bodies*	2	10
Ure	*Dictionary of Chemistry*	6	10
Black	*Lectures on Chemistry*	6	10

Table 5.2: Popular Titles in Chemistry

Reader	Charges	Reader	Charges
Dexter, Franklin	48	Wyman, Rufus	8
Tuckerman, Edward	21	Quincy, Josiah, Jr.	8
Belknap, John	20	Higginson, Francis	8
Cooke, Josiah, Sr.	18	Scholfield, Arthur	8
Hall, Joseph	17	Warren, John	8
Lawrence, Wm.	16	Boyden, Uriah A.	7
Gay, Martin	15	Cooke, Josiah, Jr.	7
Parkman, Francis	14	Andrews, Caleb	6
Bates, George	12	Parsons, William	6
Bigelow, Jacob	11	Martin, Enoch	6

Table 5.3: Top Readers in Chemistry

Figure 5.1 is the cumulative plot of chemistry charges. The two periods of elevated interest in chemistry are November 1831, to December 1832, and February 1842, to May 1843. Edward Tuckerman and John Belknap were the top readers in the first period with twenty and fourteen charges respectively. Josiah Cooke, Sr., was the top reader in the second period with sixteen charges.

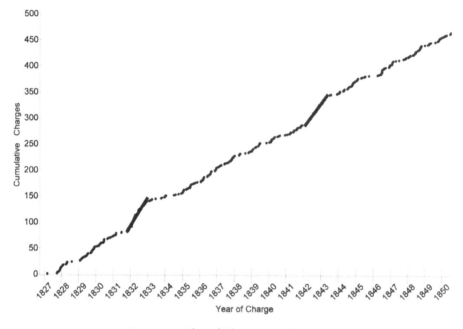

Figure 5.1: Plot of Charges in Chemistry

Period	Start	End	Readers	Charges	Interest
	Jan. 1827	Dec. 1850	130	459	52
A	Nov. 1831	Dec. 1832	19	59	150
B	Feb. 1842	May 1843	24	57	126

Table 5.4: Periods of Elevated Interest in Chemistry

Two of the most frequent borrowers of chemistry titles, Franklin Dexter and Josiah Cooke, Sr., were widely-respected Boston lawyers. The first chemistry book that each of them borrowed was Faraday's *Chemical Manipulation* [99], Dexter on July 22, 1837, and Cooke on May 5, 1842.

If you wanted to learn laboratory chemistry through hands-on laboratory experimentation, Faraday's book was the one to have open on the bench.[6] *Chemical Manipulation* is a compilation of small, focused educational experiments which enable the reader to discover—in the manner of Montessori and Pestalozzi—how things really are. Here's one of Faraday's learning experiments taken from *Chemical Manipulation*:

> Heat a globule of antimony, and when in full combustion throw it through the air against a wall; remark its combustion in the air before and after it is broken. [99, p. 605]

This mode of learning about a property of antimony may not be found in today's high school chemistry laboratories nor was it, one would suspect, seen very often in the Athenæum's demonstration room, but there is no doubting that the experiment would have made a lasting impression regarding the nature of antimony.

5.2.1   Josiah Parsons Cooke, Sr. and Jr.

Bernard Cohen credits Josiah Parsons Cooke, Jr.,[7] as being the "first university chemist to do truly distinguished work in the field of chemistry" in the United States. [56, p. 672] Biographies of Josiah Cooke, including those of memorialists William Jensen [140] and Charles Loring Jackson [138], rarely fail to mention that Cooke's interest in chemistry was sparked by attending lectures at the Lawrence Scientific School and by reading Turner's *Elements of Chemistry*. Jackson, for example, recalls:

> Young Cooke grew up a quiet boy, little given to sports out of doors, especially as early in his boyhood a course of lectures given in Boston by the elder Silliman kindled in him an enthusiasm for chemistry, which continued to blaze till the end of his life, and led him to pass all his spare time, not on the playground, but in a little laboratory which he had fitted up in his father's house. Here he attacked the science by experiment, guided by the bulky volume of Turner's *Chemistry*, and secured a mastery of the subject which would have been highly creditable with a good instructor, but without a teacher of any sort was most surprising. Yet, while a remarkably able student in chemistry and also in mathematics, he had neither taste nor aptitude for the dead languages, and it was only with much difficulty that he surmounted the barrier of Greek and Latin which guarded the approach to Harvard College. [138, p. 2]

---

[6]Gilbert chemistry sets wouldn't be available until 1917.
[7]Henceforth, 'Josiah Cooke' and 'Cooke' standing alone will refer to Josiah Cooke, Jr.

William Jensen begins his article "Physical Chemistry before Ostwald: the Textbooks of Josiah Parsons Cooke," as follows:

> [Josiah Cooke] was attracted to chemistry as a young teenager, after attending a series of Lowell Lectures on the subject given in Boston by Benjamin Silliman the elder of Yale University, and he soon constructed a "rudimentary" laboratory in the woodshed behind the family house at Winthrop Place in Boston. Here he taught himself chemistry by working through Edward Turner's massive (666 pages) text, *Elements of Chemistry*. [140, p. 10]

On June 17, 1842, about a month after Josiah Cooke, Sr., borrowed Faraday's *Chemical Manipulation*, he returned to the Athenæum and borrowed "Terner's Chemistry." Figure 5.2 shows the entry on Page 334 of Volume II recording this charge.

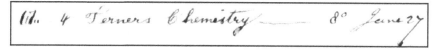

Figure 5.2: Josiah Cooke, Sr. Charge of "Terner's" Chemistry

One of the many delights of reading at a library with a long historical tail such as the Boston Athenæum is discovering that a book a father borrowed for his son 150 years ago is still on the shelf today and, as can be seen in Figure 5.3, Turner's *Chemistry* is a mere ten books to the left of a book that Cooke, Jr., would write, *Elements of Chemical Physics*.

Figure 5.3: Turner and Cooke on the Shelf at the Boston Athenæum

Josiah Cooke entered Harvard in 1843 and attended lectures in mathematics and physics given by Benjamin Peirce and Joseph Lovering. In his memoir of Cooke, Jackson refers to the undergraduate Cooke as both "a remarkable student in mathematics" and "distinguished in mathematics." [138, p.2] Table 5.5 lists titles of the mathematics books that Cooke borrowed from the Harvard library during his undergraduate years (above the line) and while he was a tutor in mathematics (below the line). Table 5.6 lists the mathematics books his father charged the Athenæum during his undergraduate years. Given his dual interest in mathematics and chemistry, it is not at all surprising that Cooke would join the search for the periodic table.

Author	Short Title	Date
Bourdon	*Applied Algebra*	Sept. 23, 1845
Boucharlat	*Théorie des Courbes*	Nov. 25, 1845
	Cours de Mathématiques	Nov. 25, 1845
Leslie	*Geometry*	Dec. 2, 1845
Hind	*Principles of Differential Calculus*	Mar. 20, 1845
Lacroix	*Calcul Différentiel et de Calcul Intégral*	May 12, 1845
Hall	*Calculus*	May 26, 1845
Hind	*Principles of Differential Calculus*	Sept. 21, 1846
Lardner	*Calculus*	Oct. 12, 1846
Hutton	*Tracts on Mathematics*	Dec. 10, 1846
	Elemens de Calcul Différential	Mar. 27, 1847
	Calculus	Mar. 27, 1847
Hall	*Differential Calculus*	Mar. 29, 1847
Loomis	*Elements of Trigonometry*	Sept. 4, 1849
Lacroix	*Calcul Différentiel et de Calcul Intégral*	Oct. 8, 1849
Simpson	*Fluxions*	Oct. 8, 1849
Legendre	*Geometrie*	Nov. 5, 1849
Whitlock	*Geometry*	Nov. 12, 1849
Woodhouse	*Trigonometry*	Nov. 12, 1849
Mayer	*Traité d'Algèbre*	Nov. 12, 1849
Leslie	*Geometry*	Nov. 19, 1849

Table 5.5: Mathematics Books Charged by Cooke, Jr., at Harvard

Author	Short Title	Date
Sherwin	*Mathematical Tables*	Aug. 28, 1845
Biot	*Géométrie Analytique*	Sept. 20, 1845
Lacroix	*Eléméntaire du Calcul Différentiel*	Jan. 17, 1846
Lacroix	*Calcul Différentiel et du Calcul Intégral*	Mar. 26, 1846
Carnot	*Calcul Infinitésimal*	Jan. 16, 1846
Lacroix	*Eléméntaire du Calcul Différentiel*	Sept. 19, 1846

Table 5.6: Mathematics Books Charged by Cooke, Sr., at the Athenæum

5.2.1.1 Moseley's *Illustrations of Mechanics*

Josiah Cooke, Sr., borrowed Moseley's *Illustrations of Mechanics* [183] on November 23, 1842, while his son was still a student at Boston Latin. Moseley[8] was a Fellow of the Royal Society and a professor of natural philosophy and astronomy at King's College, London. The book was the first in the series titled *Illustrations of Science* published by King's College. The purpose of the series was…

> …to promote this great business of PRACTICAL EDUCATION, by supplying to the instructors of youth a system of elementary science, adapted to the ordinary forms of instruction. [183, p. vii]

Moseley's book consists of 307 short educational experimental essays covering subjects from the *Tenuity of the Fibres of a Spider's Thread* (Article #17) to *A Candle fired from a Musket will Pierce through a thick Board* (Article #250). The instructions for each experiment are firmly centered on a particular physical phenomena and many of the entries included an experiment the reader could conduct to more fully grasp the article's teaching. Writes Moseley:

> The illustrations of the mechanical properties of matter and the laws of force are drawn *promiscuously*[9] and almost equally from ART and NATURE. [183, p. ix]

By 'art' Moseley means what we would call 'science,' a concern with the hidden laws of nature. By 'nature' Moseley means that which we observe. Science is a matter of gaining understanding of the hidden laws by means of experimentation. To Moseley's way of thinking, how the laws got there is not science's brief. What they are, is.

---

[8]Not to be confused with Henry G. J. Moseley who discovered the method for determining atomic number.

[9]Moseley's italics. O.E.D. "Early 17th century: from Latin *promiscuus* 'indiscriminate' (based on *miscere* 'to mix') + -ous. The early sense was 'consisting of elements mixed together' ".

While some of the experiments Moseley recommends are his own, the majority of them are drawn from the scientific literature of the day. He includes, for example, Œrsted's experiments with aëriform, Marriotte's experiments with the elasticity of air, Coulomb's torsion balance, Vicat's tables of the strength iron, Mitscherlich's crystallization of sulphur, Lussac's measurements of capillary action, Graham's pendulum, Prince Rupert's drops, Wheeler's clock, Atwood's machine, and Borda's false balance to name just a few.

In the Introduction to *Illustrations of Mechanics*, Moseley channels Lucretius:

> It is in the inaccessible minuteness of matter that the principles of the science treated of in the following pages have their origin. Matter is composed of elements, which are inappreciably and infinitely minute; and yet it is within the infinitely minute spaces which separate these elements that the greater number of forces known to use have their only sensible action. [183, p. ix]

Cooke's mechanical-universe vision of chemistry as well as his skill in the chemistry laboratory might be traced in some small part back to his reading of Mosely's do-it-yourself experiments.

5.2.1.2 Relations between Atomic Weights

In his 1854 "Numerical Relation," paper Cooke divided the fifty-eight elements known at the time into six classes, each with its own formula relating the atomic weights of the elements in the class.[10] The class he called the Five Series, for example, consisted of carbon, boron, and silicon. Referring to the atomic weight of these three elements as w_0, w_1, and w_3 respectively, Cooke's formula for the atomic weights of the elements in the Five Series[11] is

$$w_n = 6 + 5n. \tag{5.1}$$

Based on this formula, Cooke's estimate of the atomic weight of carbon was $w_0 = 6 + 5 \times 0 = 6$, that of boron was $w_1 = 6 + 5 \times 1 = 11$, and that of silicon was $w_3 = 6 + 5 \times 3 = 21$. The accepted values for these atomic weights at the time were 6, 10.9, and 21.3 respectively so the atomic weights produced by Cooke's formula for the Five Series are within experimental error of the time.

The reader undoubtedly noticed the absence of $w_2 = 6 + 5 \times 2 = 16$ from the Five Series computations. The element with atomic weight 16 is oxygen but Cooke placed oxygen in two other series, the Nine Series and the Six Series based on its similarities with other elements in those two series.

---

[10] In the first half of the nineteenth century atomic weights were by default relative to hydrogen.

[11] The series is called the Five Series due to the multiplier 5 in its characterizing equation.

Cooke's emergent periodic table was, like all the others at the time, a work in progress. He fastened on a numerical relationship among measured values, used the relationship to organize the elements, and then tested the resulting quantitative organization against the elements' qualitative properties. What made the search for the periodic table such a siren song was that not only were the numbers he had only approximate but he—and everyone else—knew that there missing numbers; viz., the atomic weights of undiscovered elements. And yet the numbers seemed to fit together so nicely, almost.

Cooke's purpose in developing his series had as much to do with pedagogy as it did with peering into the atom. As he put it in the "Numerical Relation" paper:

> Every teacher of Chemistry must have felt the want of some system of classification like those which so greatly facilitate the acquisition of the natural-history sciences....The object of the classification was simply to facilitate the acquisition of Chemistry, by bringing together such elements as were allied in their chemical relations considered collectively. [67, p. 236]

In his survey of the history of the periodic table [104], Jack Fergusson includes Table 5.7 which shows Cooke's position in the race for the table.

Date	Person	Contribution	Elements
Early Groupings			
1789	Lavoisier	Metals/nonmetals	23
1817	Döbereiner	Triads of elements	45
1827	Gmelin	Triads of elements	50
1843	Gmelin	Triads of elements	54
1850	Pettinkofer	Equivalent weights	58
1851	Dumas	Triads and bigger groups	58
1852	Kremers	Mathematical series	58
1853	Gladstone	Triads, similar atomic weights	58
1854	**Cooke**	**Groups of six**	**58**
1857	Lenssen	Triads of elements	58
1857	Odling	Triads and bigger groups	58
1858	Mercer	Parallels in groups	58
1660	Carey Lea	Based around at Wt =45	58
Periodic System			
1862	de Chancourtois	3D helical representation	61
1863	Newlands	Law of octaves	62
1864	Odling	Periodic system	62
1866	Hinrichs	Atomic weight relationships	62
1868	Meyer	Periodic system	63
1868	Mendeleev	Periodic system	63

Table 5.7: Timeline of the Development of the Periodic Table

One of Cooke's laboratory assistants, a man who would go on to become a lion of antebellum mathematics, Benjamin Peirce, hailed Cooke's "Numerical Relation" paper as a "wonderful discovery." [138, pp. 5–6]. And one of Cooke's students, Theodore Richards, would go on to win America's first Nobel laureate in chemistry.

5.2.1.3 Sylvester's Combinatorics

On his own Athenæum account, Josiah Cooke, Jr., charged Volume 24 of Tilloch's *Philosophical Magazine* three times: March 10, 1845; April 1, 1846; and March 30, 1848. While there can be no certainty as to which article in the volume interested him or even if it was the same article on all three occasions, one may hypothesize.

For openers, there are a number of articles in Volume 24 reporting on the determination of the atomic weight of various substances: Hofman's "A Chemical Investigation of the Organic Bases contained in Coal-Gas Naptha" and Stenhouse "On the Products of the Distillation of Meconic Acid," to name two. These certainly would have interested Cooke yet one imagines that he would have jotted down such late-breaking atomic weight values while reading at the Athenæum rather than taking the volume home.

On the assumption that it was the same article all three times and that it was long enough to justify borrowing the volume, is there an article in Volume 24 that Cooke might have come back to either as a reference or because it was difficult but at the same time particularly relevant to his work? My candidate is a dense, 12-page article by the British mathematician J. J. Sylvester, "Elementary researches in the Analysis of Combinatorial Aggregation."[12] [227].

Cooke was in the throes of understanding the attempts of other chemists to convert patterns of atomic weights to mathematical formulas as well as developing his own. As of 1845, when he borrowed the volume for the first time, fifty-seven elements had been discovered. Fitting equations to their atomic weights was much more than just drawing a line through a scatter of points. It was more like a game of Sudoku where the relative atomic weights of various elements had to come out the same regardless of which compounds were broken down to determine them and they had to go back together to make other compounds which when broken down…well, you get the picture.

Sylvester's paper on the mathematics of combinatorial aggregation is very much in the style and spirit of the mathematics of Charles S. Peirce who may well have suggested it to Cooke. An example of the problems that Sylvester set for himself is the following:

> Given four elements,[13] say a, b, c, and d, find all ways of arranging them in pairs so that each element appears once and only once in each arrangement.

The arrangement $\{(ab), (ac)\}$ doesn't qualify because a appears twice and d not at all. There are just three arrangements that solve Sylvester's puzzler:

$$\{(ab), (cd)\}$$
$$\{(ac), (bd)\}$$
$$\{(ad), (bc)\}$$

---

[12] The article ends with "[To be continued]" but there is no continuation in either Tilloch's magazine or Sylvester's collected works.

[13] Sylvester calls them elements in his paper but it is highly unlikely that he was thinking of chemical elements.

Sylvester considers all sorts of variations of his problem—six elements, ten elements, arrangement in triples, arrangement in quadruples, each element appears twice, each element appears three times, and so forth. The contribution of Sylvester's paper is an algorithm for constructing qualifying arrangements given various such conditions on the arrangements.

Sylvester introduced his study of combinatorial aggregation as follows:

> The ensuing inquiries will be found to relate to combination-systems, that is, to combinations viewed in an aggregative capacity, whose species being given, we shall have to discover rules for ranging or evolving them in classes amenable to certain prescribed conditions. [227, p. 285]

Reading "chemical elements" for "species," "chemical compounds" for "classes," and "chemical formulae" for "rules," it is not at all difficult to understand why Cooke might have repeatedly taken the volume home for careful study.

5.2.1.4 Proust to Prout to Turner

The idea of using mathematical expressions to construct laws about chemical properties emerged at the end of the eighteenth century in tandem with revolutionary notions of chemical building blocks and chemical compounds. Two keys were needed to open the door behind which was the path leading to the periodic table of elements. One key was contributed by a French chemist, Joseph Louis Proust, and the other by an English chemist, William Prout.

Proust's key unlocks chemical compounds. It is called the Law of Definite Proportions and says that a chemical compound always contains exactly the same proportion of elements by mass. No matter where you fill a canteen with water, the ratio of the mass of hydrogen in your canteen to the mass of oxygen in your canteen will always be the same.

Prout's key unlocks the building blocks. Prout noticed that the atomic weights of all the elements known in the early eighteenth century were integer multiples of the atomic weight of hydrogen and conjectured that hydrogen was the Lego block of physical chemistry. As it turned out, Prout's conjecture was proven false by Edward Turner but its flotation had the redeeming value of focusing attention on the arithmetic regularity of atomic weights.

Proust says that a given compound always consists of the same proportion of elements. Prout says that properties of elements such as atomic weight are mathematically related. And Dalton contributes a physical model, the atom, such that "Atoms of a given element are identical in size, mass, and other properties; atoms of different elements differ in size, mass, and other properties."[14] It's mathematics all the way down, but where is the master equation?

---

[14]Quoted from the Wikipedia entry for John Dalton. The notion of an atom has a history that predates

Section II of Turner's *Elements of Chemistry* is titled "On the Proportions in which Bodies Unite, and on the Laws of Combination." In introducing the section, Turner sets forth the three laws:

> The combination of substances that unite in a few proportions only, is regulated by three remarkable laws.

> The first of these laws is, that the composition of bodies is fixed and invariable; that a compound substance, so long as it retains its characteristic properties, must always consist of the same elements united together in the same proportion.[15]

> The second law of combination is still more remarkable than the first. It has given plausibility to an ingenious hypothesis concerning the ultimate particles of matter, called [Dalton's] *atomic theory*....[The law] may be stated in the following terms. When two substances, A and B, unite chemically in two or more proportions, the numbers representing the quantities of B combined with the same quantity of A are in the ratio of 1, 2, 3, 4, &c.; that is, they are multiples by some whole number of the smallest quantity of B with which A can unite.

> The third law of combination is fully as remarkable as the preceding, and it is intimately connected with it, that bodies unite according to proportional numbers; and hence has arisen the use of certain terms as Proportion, Combining Proportion, or Equivalent, to express them. [240, pp. 90–93]

At the beginning of the following section, "On the Atomic Theory of Mr. Dalton," Turner states in no uncertain words that he is writing a Boylean experimental essay:

> The brief sketch which has been given of the laws of combination will, I trust, set in its true light the importance of that department of chemical science. It is founded on experiment alone, and the laws which have been stated are the mere expression of fact. [240, p. 167]

Turner's *Chemistry* is filled with thumbnail analyses of the following sort:

> [I]t appears that there are but two oxides of copper, and that they are thus constituted

---

Dalton by many centuries. It is often credited to the fifth-century BC Greek philosophers, Democritus and Leucippus. Atoms are also found in Lucretius' *De rerum natura*: "The atoms must a little swerve at times—". See Stephen Greenblatt, *The Swerve: How the World Became Modern* [121].

[15]Turner tips his hat to Proust and presciently says in passing that this law, if true, "would shake the whole science of chemistry to its foundation."

	Copper		Oxygen		
Protoxide	64	+	8	=	72
Peroxide	64	+	16	=	80

Table 5.8: Estimating the Atomic Weight of Copper

Consequently, if the first be regarded as a compound of one atom of each element, 64 is the atomic weight of copper.

Given Cooke's early and abiding enthusiasm for mathematics, it seems to me that his career as well as his accomplishments can not only be traced back to Turner's *Elements of Chemistry* but to this very page in Turner's book.

5.2.2 Franklin and William Dexter

Franklin Dexter was "one of the giants of the Boston bar in the early 19th century." [204]. He was the son of Samuel Dexter, another renowned Boston barrister. He graduated Harvard in 1812, and immediately thereafter studied law with Samuel Hubbard. In addition to being a member of both the Massachusetts Senate and House of Representatives from time to time, he loomed large in numerous headline-grabbing court cases including Benjamin Whitman's libel suit against William Snelling, the editor of the *New England Galaxy*; Daniel Webster's libel suit against Theodore Lyman;[16] the embezzlement trial of the president of the Phoenix Bank of Charlestown, William Wyman;[17] and the talk-of-the-town murder trial of Mrs. Hannah Kinney.

How is it that this busy member of the Boston bar charges forty-eight chemistry titles between July of 1837 and March of 1850? Another father and son relationship provides the explanation. During the period that Dexter *pere* was borrowing the Athenæum's chemistry books, Dexter *fils* was at the beginning of his life work in chemistry.

William Prescott Dexter[18] was born at Boston, on the 10th of December, 1820. He studied medicine at Harvard College, and graduated in 1838.[19] After leaving college, he practised medicine at Brookline for a few years, but his strong interest in pure science and his longing to devote himself solely to scientific pursuits made him feel the irksomeness of the ordinary

---

[16] Another of our readers.

[17] ...not known to be related to Rufus or Morrill.

[18] Henceforth 'Dexter' will refer to Franklin's son, William Prescott Dexter.

[19] William Dexter was the second cousin three times removed of Aaron Dexter, Chair of Chemistry and Materia Medica at Harvard from 1783 to 1816 and Professor Emeritus until his death in 1829.

routine of a medical practitioner's life, and induced him to give up the medical profession, and to go abroad in order to study chemistry. From this time to the end of his life he devoted himself entirely to the prosecution of that science. [206, p. 364]

Precocious is putting it mildly. Prepared at Boston Latin, he entered Harvard College at the age of thirteen, graduating eight years later as a Doctor of Medicine. Dexter's reading during his years at Harvard was primarily in history and medicine although he did charge Webster's *Manual of Chemistry* in his junior year and Thomson's *Chemistry* twice in his senior year.[20]

Figure 5.4 is the cumulative charges plot of Franklin Dexter's chemistry charges which we are taking to be on behalf of his son. Table 5.9 lists the details of the three periods of elevated interest in the subject.

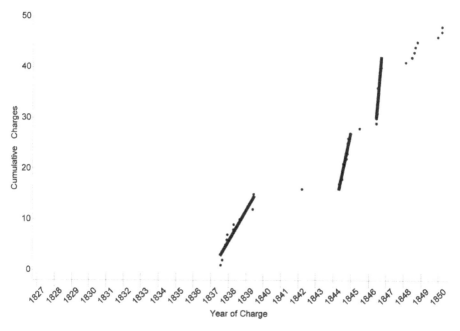

Figure 5.4: Plot of Franklin Dexter's Charges in Chemistry

---

[20] The same book by Thomson that Tredgold used as a reference for his effect of fuel numbers.

Period	Start	End	Charges	Interest
A	July 1837	June 1839	15	16
B	Apr. 1844	Dec. 1844	11	46
C	June 1846	Aug. 1846	12	118

Table 5.9: Periods of Franklin Dexter's (Apparent) Interest in Chemistry

Twenty-two of Franklin Dexter's forty-eight charges of chemistry titles were for the journal *Annales de Chimie*. Twelve of these charges were for more than one volume. If our supposition regarding Franklin's charges is correct, William read his way through sixty-four volumes of the *Annales*, devouring forty-two volumes in the summer of 1846 alone.

5.2.2.1   William Dexter's *Tabulæ Atomicæ*

In 1850 William Dexter published *Tabulæ Atomicæ: The Chemical Tables for the Calculation of Quantitative Analyses of H. Rose*[72], a supplement to Rose's popular *Manual of Analytical Chemistry*. Like Bowditch and many other mathematical table makers before him, Dexter realized that publishing a book filled with tables of numbers entailed careful, if not to say mind-numbing, proofreading:

> As the value of a work of this kind depends entirely upon its correctness, I may state that every calculation was performed by myself, both by direct division and by the use of logarithms. The columns of multiples were computed separately by myself and another, and our results compared both before and after they were transcribed. Finally, to avoid as far as possible errors of the press, each sheet, besides the usual correction, has been most carefully revised by myself. [72, pp. iii–iv]

The primary contribution of Dexter's book was to gather together in one place all the atomic weight values that were scattered throughout the literature of the day and to organize them into a easily used format. If you had 31.35 grains of silver chloride and you wanted to know how much silver your sample contained, then entering Table III Argentum and scanning down to the fourth line, Chloridum Argenticum, you would find that you simply had to multiply the amount silver oxide you had by 0.75276 to get the amount of silver it contained: $31.35 \times 0.75276 = 23.59912$ grains.

Dexter goes to great lengths to perform this multiplication by providing a table single-digit multiples of 0.75276, so you can perform the multiplication by table lookup and addition as follows:

$$
\begin{aligned}
31.35 \times 0.75276 = \quad & 10 && \times(3 \times 0.75276) \\
+ \; & 1 && \times(1 \times 0.75276) \\
+ \; & 0.1 && \times(3 \times 0.75276) \\
+ \; & 0.01 && \times(5 \times 0.75276) \\
= \; & 23.59912
\end{aligned}
\tag{5.2}
$$

The values within the parentheses being taken out of Dexter's table single-digit multiples of 0.75276.

If you didn't want to use his tables of multiples, Dexter also supplies the logarithm of his multipliers so that after you compute the logarithm of 31.35 you just have to add it to the logarithm of 0.75276 and take the anti-logarithm of the result to get the number you seek.

$$
\begin{aligned}
\log 31.35 + \log 0.75276 &\approx 1.4962375 + (-0.1233435) \\
&= 1.3728940 \\
&\approx \log 23.59912
\end{aligned}
$$

Dexter recommends the method using logarithms over his multiplier tables, because it is…

> …free from all chance of error arising from a wrong placing of the decimal point; an error which, by the other method, may easily be committed. [72, p. xi]

which can't be denied but it does lead one to wonder if chemists of the day had some difficulty with multiplication and yet found taking logs and anti-logs a breeze.

There is a curious footnote in Dexter's discussion of the use of his multiplier tables:

> I would add, that Rose considers logarithms as not leading to sufficiently accurate results for the calculation of exact analyses. This is true if the logarithm be carried only to five decimal places; but if tables of seven decimals be made use of, the results will be found more exact than if made with Rose's Tables. [*Ibid.*]

Rose recommended that chemists use 5-place multiplier tables and throw away the fifth place to do their multiplications. Dexter recommended that chemists use 7-place logarithm tables. And all of this while working with atomic weight measurements for which there was scant consensus on even the second place.

5.3   Purposeful Reading about Botany

Edward Tuckerman was a respected Boston lawyer much like Josiah Cooke, Sr., and Franklin Dexter. Unlike his colleagues however, the scientific and technical books Tuckerman borrowed we can readily surmise were for his own use.

Tuckerman was one of the widest read of the Athenæum's readers in experimental philosophy, logging eighty-two charges against forty-five titles by forty authors across eight of the subject matter categories. Table 5.10 tabulates Tuckerman's titles and charges across these categories.

Domain	Titles	Charges
Botany	13	25
Chemistry	16	21
Conchology	5	17
Mineralogy	3	7
Zoology	3	6
Geology	3	4
Ornithology	1	1
Entomology	1	1

Table 5.10: Categories of Edward Tuckerman's Charges

Tuckerman was the author of numerous articles and tracts on lichens. His capstone two-volume work, *A Synopsis of the North American Lichens* [239], published in 1882 (Volume I) and 1888 (Volume II) is in the Athenæum's collection. The descriptions of lichens in Tuckerman's synopsis are a marvel of attention to detail:[21]

> 34. B. parasitica, (Fl.) Th. Fr.; thallus foreign; apothecia minute, sessile, flat, or at length a little convex, naked, with a regular, thin margin, the hypothecium blackish-brown. Spores ellipsoid and oblong, 4-locular, 9-18 by 3-6 mic.—*Lecidea, Nyl. Prodr. p. 144; Lich. Par. n. 68. Dactylospora, Koerb. Syst. p. 271.* [239, p. 107]

Experimental essays for identifying plants are not in their manner of discourse any different than experimental essays for identifying chemical compounds or minerals. Here is such an assay essay from the most popular book on botany, Loudon's *Encyclopedia of Plants*:

---

[21]See the passage on lichenologists at the Smithsonian in Bill Bryson's *A Short History of Nearly Everything* [47] for more insight into lichenologists and detail.

M. paradisiaca rises with a soft herbaceous stalk fifteen or twenty feet high, with leaves often more than six feet long, and near two feet broad. When the plant is full grown, the spike of flowers appears from the centre of the leaves; it is near four feet in length, and nods on one side. The fruit which succeeds the fertile flowers on the lower part of the spike is eight or nine inches long, and above an inch in diameter, a little incurved, with three angles; at first green, but when ripe of a pale yellow color. The skin is tough, and within is a soft pulp of a luscious sweet flavor. The spikes of fruit are often so large as to weigh upwards of forty pounds. [167, p. 244]

Figure 5.5 is the cumulative charge plot for botany. There are no obvious periods of elevated interest.[22]

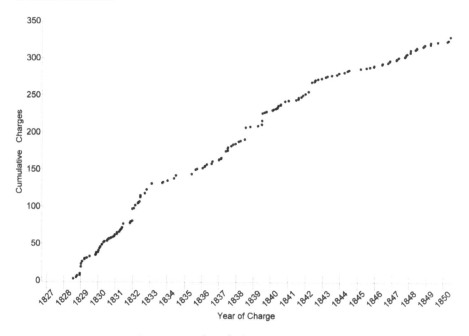

Figure 5.5: Plot of Charges in Botany

One hundred and six readers generated 335 charges against seventy-one titles in botany. Charges for the most popular titles are listed in Table 5.11. The top three readers were John W. Boott, George B. Emerson, and Edward Tuckerman, the first with twenty-six charges and the later two with twenty-five charges each.

---

[22]The cumulative charge curve is approximated nicely by $5\sqrt{d} - 120$ but this only says that interest in botany is falling off which is obvious from the plot.

Author	Short Title	Readers	Charges
Curtis	*Botanical Magazine*	13	50
Loudon	*Encyclopedia of Plants*	14	29
Bigelow	*American Medical Botany*	17	25
Sowerby	*English Botany*	7	22
Bigelow	*Florula Bostoniensis*	15	19
Candolle	*Philosophy of Plants*	11	13
Torrey	*Flora of the United States*	8	12
Smith	*Physiological and Systematical Botany*	10	12
Nuttall	*Genera of North American Plants*	9	12

Table 5.11: Popular Titles in Botany

Not listed in Table 5.11 is Hitchcock's *Report on the Geology, Mineralogy, Botany, and Zoology of Massachusetts* which was charged twenty-five times. Containing as it does four of the subject matter domains one can't be certain which one might have attracted the reader. We will return to this title in Chapter 9.

Three readers in experimental philosophy at the Boston Athenæum had species of plants named after them. Their names and flora namesakes are listed in Table 5.12.

Reader	Latin Name	Common Name
Jacob Bigelow	*Carex bigloweii*	Bigelow's sedge
John W. Boott	*Prenanthes boottii*	Boott's rattlesnake root
Edward Tuckerman	*Carex tuckermanii*	Tuckerman's sedge

Table 5.12: Plants Named after Athenæum Readers

Jacob Bigelow's name was also bestowed on a genus of flowering plants, *Bigelovia*,[23] by de Candolle to whom Bigelow sent specimens. The botanical interests of the third top reader, George B. Emerson, leaned more toward trees. Emerson is credited with being the individual who brought the Arnold Arboretum[24] to fruition. [132]

---

[23] ...(or Bigelowia) in the family *Rubiaceae* (or *Compositae*) in the major group *Angiosperms*; known better to non-botanists as goldenrod.

[24] James Arnold, after whom the arboretum is named, was Emerson's brother-in-law.

5.3.1 Jacob Bigelow's *American Medical Botany*

In his most widely-quoted medical paper, "On Self-Limited Diseases,"[25] delivered to the Massachusetts Medical Society on May 27, 1835, Jacob Bigelow describes the distinct nature of experimental philosophy as found in the practice of medicine:

> In the study of experimental philosophy, we rarely admit a conclusion to be true, until its opposite has been proven to be untrue. But in medicine we are often obliged to be content to accept as evidence the results of cases, which have been finished under treatment, because we have not the opportunity to know how far these results would have been different, had the cases been left to themselves. [14, 29–30]

Bigelow's view of the goal of experimental philosophy is stronger than Boyle's. Boyle sought a conclusion that convinced the reader that "This is just how things are." Bigelow sets his sights higher: "This is the only way things can be." Bigelow does, however, go on to admit that medicine by its intrinsic nature has a hard time clearing this higher bar. Throughout Bigelow's writing, both on technology and on medicine, one senses the same conviction that one finds in Cooke's writing; namely, that there were quantitative relationships underpinning all qualitative phenomena lurking in the numbers.

Bigelow was an advocate of what he called practical medicine. In "On the Treatment of Disease," a lecture he delivered some eighteen years later, Bigelow reflects:

> Preëminent among the inexact and speculative sciences stands *practical medicine*,[26] a science older than civilization but still unsettled in its principles. [15, p. 7]

The scientific rigor of blind trials and the oversight of regulatory bureaucracies had not been brought to bear on the evaluation effectiveness of medicines when Bigelow published *American Medical Botany* between 1817 and 1821. Each of the three volumes of the work contains twenty experimental essays, each one of which is about a particular plant and all of which follow the same three-fold organization. The first section is a full-page color drawing of the plant. The second section, the bulk of the essay, is a three-part discourse on the nature and medical use of the plant. The third section consists of botanical and medical references relevant to the plant.

The opening drawing of all but three of the sixty essays was drawn by Bigelow himself. In the first one hundred impressions of Volume I, these drawings were hand-colored but Bigelow objected to having to put paint-by-number black lines around the

---

[25] Self-limited diseases are those which "may tend to death, or recovery, but are not known to be shortened, or greatly changed, by medical treatment."

[26] Bigelow's emphasis.

regions to be colored on the grounds that these black lines did not exist in nature. For the remaining drawings in Volume I[27] and for all the drawings in Volumes II and III, Bigelow used aqua-tint printing to color the drawings.[28]

Three-part discourse comprising the second section consists of: 1) a botanical description of the plant, 2) an enumeration of the the physical and chemical qualities of the plant, and 3) a discussion of preparation and use of the plant as a medicine. The third part of this section occasionally included the description of situations in which contact with the plant might lead to consultation with a medical practitioner. In a small number of the essays—hemlock, ginseng, and tobacco, for example—Bigelow includes an extensive history of the medicinal aspects of the plant.

5.3.1.1 *Rhus vernix*[29]

Bigelows's essay on *Rhus vernix* starts with the following:

> The fine, smooth foliage of the *Rhus vernix* renders it one of the most elegant of our native shrubs, while its well known poisonous qualities make it an object of aversion, and deter most persons from a new inspection of its structure and characteristics. [13, p. 94]

This introduction is followed by botanical details: "wood light and brittle, and contains much pith," "leaves are pinate, the leafets oblong or oval," "the calyx has five ovate segments," and so forth. The botanical description continues:

> [The] juice has an opaque, whitish appearance, and a strong, penetrating, disagreeable smell. On exposure to the atmosphere, its colour soon changes to a deep black. It is extremely slow in drying, and permanently retains its black color. [13, p. 99]

On a local field trip in October of 1814 Bigelow reported that he "collected several ounces of this juice from a thicket of trees in Brighton." On returning to his laboratory he submitted his sample to chemical examination including boiling it in water, dissolving it in alcohol, and combining it with sulphuric acid. He concludes the chemical analysis section with:

---

[27]Until recently it was thought that there were only these two states of Volume I, one state colored by hand and the other state colored by printing. In the copy of Volume I in the collection of the Athenæum—a copy donated by Bigelow himself—the first ten plates are colored by hand and the remaining ten plates are colored using the aqua-tint printing process.

[28]In describing *American Medical Botany*, Bigelow's memorialist in the *Proceedings of the American Academy of Arts and Sciences*, for example, states that the drawings were printed with plates "colored by a new process of his own invention." Bigelow was well abreast of technology of the day as witnessed by his book, *Elements of Technology*. It is highly unlikely that Bigelow was unaware of aqua-tint printing and its current use in Europe.

[29]Also known as poison sumach and dogwood.

> The chemical constitution of the juice of the *Rhus vernix* seems, from the foregoing experiments, to be most analogous to that of the balsams, consisting chiefly of a resin and an essential oil. [13, p. 102]

There is no distinct differentiation between the physical and chemical descriptions. In almost all cases, the descriptions of both aspects of the plant are qualitative and confined to attributes that would be relevant to preparing the plant as a medicine. The medical section of *Rhus vernix* is introduced by:

> A very disturbing, cutaneous disease, it is well known, ensues in many persons from the contact, and even from the effluvium of this shrub. [*Ibid.*]

and goes on to describe the effect of *Rhus vernix* in graphic detail: "sense of itching," "tumefaction of the hands and face," painful burning sensation," "purulent appearance," "pustules," "yellowish incrustation," etc.

In addition to reporting on his own experiences, both with the plant itself and with its use in various medical situations, Bigelow reports on a field report of one of his colleagues:

> Kalm, in his travels in North America, mentions a person who, by the simple exhalation of the *Rhus vernix,* was swollen to such a degree, that "he was stiff as a log of wood, and could only be turned about in sheets." [13, p. 104]

Bigelow also quotes at length from a letter written to him by the gentleman who had accompanied him on the field trip to Brighton, Dr. A. I. Pierson. Bigelow notes that while he himself had "experienced no ill consequences, except a slight vesicular eruption on the backs of the hands and about the eyes," such was not the fate of Dr. Pierson. Pierson's letter, written in July of 1815, over nine months after the field trip ends with

> I am still subject to an eruption of watery pustules between my fingers, which dry up, and the cuticle peels off. [13, p. 107]

Bigelow covers the treatment of exposure to *Rhus vernix* in the following entry, *Rhus radicans,* poison ivy. Recommendations include opium, purging with neutral salts, blood-letting, and the external application of acetate of lead or "a solution of corrosive sublimate." Rather startlingly, at least from the vantage point of the twenty-first century, Bigelow also reports that preparations of poison ivy have been administered internally for pulmonary consumption and externally for "obstinate herpetic eruption" but cautions that...

> ...the plant under consideration is too uncertain and hazardous to be employed in medicine, or kept in apothecaries' shops. [13, p. 30]

But I am content, provided experimental learning be really promoted, to contribute even in the least plausible way to the advancement of it; and had rather not only be an under-builder, but even dig in the quarries for materials towards so useful a structure, as a solid body of natural philosophy, than not do something towards the erection of it.

A Proëmical Essay
Robert Boyle

Chapter 6

Geology, Mineralogy, and Anthracite Iron

6.1 Colonel Perkins Goes Prospecting

The Boston's antebellum venture capitalists are known for their investments in textile mills and railroads. Less well known is their considerable interest in mining ventures.

In 1809 Athenæum benefactor, Thomas H. Perkins, purchased a one-third interest in a long-abandoned lead mine[1] in Northhampton, Massachusetts for $2,666.66. Returns from an exploratory shaft had been promising, showing commercial quantities of lead and silver as well as traces of zinc, copper, and barite. [207, p. 216] Perkins hired Yale's Benjamin Silliman to visit the mine and write a report on its prospects. Silliman's report appears in Volume 1 of *American Mineralogical Journal*. [216]

The Athenæum's copy of the *American Mineralogical Journal* was borrowed only once and that time by Francis Alger whom we will meet below. Silliman's report on his visit to Perkin's mine starts with a short letter to the journal's editor, Archibald Bruce, giving some background of his field trip and apologizing for not rewriting the report for publication in a scientific journal:

> To Dr. Bruce
> *Dear Sir,*
> In the month of May last, at the request of the proprietors, I visited the Lead-Mine near Northampton, and the following account was drawn up

---

[1] In 1812 Perkins would be detained for three weeks in Morlaix, France, on suspicion of harboring anti-Jacobian sentiments. Never one to neglect his business interests, he took the opportunity to tour local lead mines. [53, p. 211].

as a report, for their use. It is now, with their consent, forwarded to you, and you have mine to insert it in your Journal, if it comes within your design. It would have been written differently, in some respects, had it been originally intended for a scientific journal, but I have not now time to give it a new form. [216, p. 63]

The report must have warmed Perkins' pocketbook. Here are some snippets from Silliman's article:

- "The vein, including every thing it contains, is a very magnificent one."
- "The quartz has shot into numerous crystals, usually very regular, sometimes large, and often so beautiful and brilliant that the cavities look as if studded over with gems."
- "Many of [the crystals] are sufficiently perfect and beautiful to deserve a place in the choicest cabinets on the earth"
- "Were the vein situated in Derbyshire...there can be no doubt that the enterprise would be vigorously pursued;"
- "This vein is the most interesting one that has ever been opened in this country."

In a footnote, Silliman informs the reader that Bruce also visited the site but he doesn't say whether or not it was at Perkins' invitation. If it was, then Perkins hired not one but two of the most eminent mineralogists in the country to appraise his mine.

Whether or not his pocketbook was warmed, Perkins was understandably quite satisfied with Silliman's report. In his biography of Silliman Chandos Brown provides a financial detail:

Perkins and Davis appeared satisfied and, to Silliman's surprise, presented him with fifty dollars in gold—he had expected at most ten dollars. [46, p. 258]

Perkins and his partners in the venture, Perkins Nichols, Isaac Davis, and David Hinkley, took Silliman's advice, hired Luther Work as the chief miner. Operations at the mine commenced in the spring of 1809. In their history of Northampton, James Trumbull and Seth Pomeroy conclude the sad tale of Perkins' lead mine adventure as follows:

[Luther Work] labored alone for about thirteen years, with no other company than his wife, but finally died of fever contracted by the foul air of the mine. [236, p. 367]

Perkins et al. abandoned the venture 1828.

At the same time that Perkins was taking a flyer on the lead mine in Northhampton, he was also nursing a 1807 investment in another mining venture, the Monkton Iron Company of Vergennes, Vermont. The business plan for Monkton Iron was another of Perkins' signature vertical integrations.

The plans for the company were on an extensive scale, almost unprece-
dented. They proposed a completely unified operation: a single company
that mined the ore, smelted it, provided the coal from their own lands for
the smelting, had the sloops to transport the ore and coal, refined the pig
iron, wrought it into nails, pots, kettles, sheets, bars, provided boarding
accommodations for the many laborers involved, and even contemplated
furnishing library books to them! It was unique for its time, and incorporated elements that a dozen years later would be hailed as pioneering
efforts when introduced at the textile mills in Lowell. [207, p. 202]

Not only does Volume I of *American Mineralogical Journal* contain Silliman's
report on the Northhampton lead mine but a mere ten pages on there is an article
about the Monkton Iron Works by gentleman mineralogist and long-time Silliman
benefactor, Colonel George Gibbs of Newport.[2]

After singing the praises of America's emerging metal working industry, Gibbs
tips his hat to the men from Boston:

> One of the principal undertakings in New-England, is the establishment
> of the Iron Works at Vergennes, in Vermont. The purchase of a bed of
> iron ore at Monckton, [*sic*] seven miles distant, led a company of gentle-
> men to erect, on the falls of Otter-Creek, the most extensive iron works
> to be found in the Northern States. Messrs. Welles, Perkins, Higginson,
> and Bradbury,[3] of Boston, have engaged in this enterprise with a liberality,
> and to an extent which, I think, must ensure them success, and certainly
> entitles them to applause. [215, p. 81]

and finishes his article on distinctly positive note:

> To conclude, the iron-works of Vergennes promise to become as impor-
> tant as any in America, and offer every encouragement to the persever-
> ance and enterprise of the proprietors. [215, p. 83]

In 1810, Bradbury, now the on-site manager of the furnace, beseeched his fellow
investors back in Boston for some reading for practical purposes:

---

[2]Col. Gibbs and Col. Perkins were at least acquaintances. In a postscript to an 1819 article in *Silliman's
Journal*, "Observations on Dry Rot," Gibbs writes "Since the above was written, I have received from Col.
Perkins, of Boston, some valuable information on the subject…" [110, p. 117] Perkins' interest in mining
ventures might be due at least in part to Gibbs. For a fleeting moment Gibbs' renowned mineral cabinet was
to be deposited at the Boston Athenæum. Not surprisingly (and perhaps thankfully from the Athenæum's
perspective), the collection ended up at Silliman's institution, Yale University.

[3]The partners in the venture were Thomas H. and his brother, James, Stephen Higginson and son
George, William Parsons, Frank Bradbury, and Benjamin Welles.

I have no books that treat on [refining] and nobody of any information here on the business. If you find any treatise on the refining of iron and can send it to me, it may assist me. [207, p. 215]

Sadly, it would be thirty years before the Athenæum would add Walter Johnson's *Notes on the Use of Anthracite in the Manufacture of Iron* [141] to its collection.

As it turned out, the only part of the business plan forMonkton Iron that showed a reliable profit was the company store. The business limped along for a quarter of a century before being sold to a Canadian, John D. Ward, in 1831. [207, p. 421]

There are no marks on the Athenæum's copy of Volume I of *American Mineralogical Journal* but it is tempting to imagine that it was once browsed by Colonel Perkins himself.

6.2   Purposeful Reading about Geology

There are five books mentioned in the books borrowed registers that have both 'geology' and 'mineralogy' in their title:

- Cleaveland's *Treatise on Mineralogy and Geology*
- Buckland's *Geology and Mineralogy Considered*
- Dana's *Mineralogy and Geology of Boston*
- Alger & Jackson's *Mineralogy and Geology of Nova Scotia*
- Hitchcock's *Report on the Geology, Mineralogy, Botany, and Zoology of Massachusetts*

The books by Cleaveland and Dana are primarily mineralogy books and have been assigned to that category. The books by Buckland and Alger & Jackson are included in the geology category. Hitchcock's report is discussed in Chapter 9 and is not included in either geology or mineralogy tabulations.

There are two books by Charles Lyell in the geology category, *Principles of Geology* and *Elements of Geology*. In many cases the entry in the books borrowed register was simply "Lyell Geology," not differentiating between the two. For the purpose of the following tabulations they have been combined under the short title *Geology*.

One hundred and sixty-seven readers generated 427 charges against forty-five geology titles. Figure 6.1 is the cumulative charge plot for geology and Table 6.1 describes the single period of slightly elevated interest. Table 6.2 lists the titles in geology with more than ten charges and Table 6.3 lists the top readers.

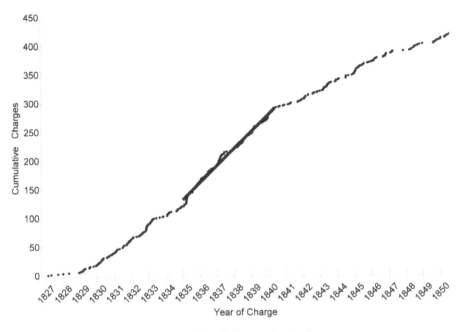

Figure 6.1: Plot of Charges in Geology

Period	Start	End	Readers	Charges	Interest
	Jan. 1827	Dec. 1850	165	427	49
A	Jan. 1835	May. 1840	70	153	80

Table 6.1: Periods of Elevated Interest in Geology

Author	Short Title	Readers	Charges
Lyell	*Geology*	54	88
Bakewell	*Introduction to Geology*	28	33
Cuvier	*Essay on the Theory of the Earth*	25	32
Greenough	*Transactions Geological Society*	9	26
Buckland	*Geology and Mineralogy Considered*	21	26
Ansted	*Geology,…Descriptive, and Practical*	18	23
Penn	*Conversations on Geology*	18	19
Beche	*How To Observe Geology*	16	16
Ure	*New System of Geology*	11	13
Mantell	*Wonders of Geology*	11	13
Cuvier	*Recherches sur les Ossemens Fossiles*	8	12
Cuvier	*Revolutions of the Surface of the Globe*	11	11
Parkinson	*Organic Remains*	8	11
Macculloch	*System of Geology*	9	10

Table 6.2: Popular Titles in Geology

Reader	Charges	Reader	Charges
Martin, Enoch	19	Belknap, John	6
Codman, Henry	16	Heard, John, Jr.	6
Randall, John	9	Alger, Cyrus	6
Parsons, William	8	Jackson, Charles T.	6
Hooper, Samuel	8	Jenks, William	6
Phillips, Jonathan	7	Bartol, Cyrus Augustus	6
Ripley, George	7	Parkman, Francis	6
Lyman, William	7	Winslow, Isaac	6
Emerson, George B.	7	Gardiner, Chandler L., Jr.	5
Cooke, Josiah Parsons, Jr.	7	Sherwin, Thomas	5
Lawrence, William Richards	7	Sullivan, Richard	5
Ware, John	7	Warren, A., Mrs.	5

Table 6.3: Top Readers in Geology

6.2.1 Alger & Jackson vs. Abraham Gesner

In 1826 Boston ironmaster and Athenæum reader, Cyrus Alger, took his teenage son, Francis, along with him when he traveled to the British colony of Nova Scotia to see about building a smelting furnace in the Annapolis basin. While his father was conducting business, Francis collected mineral specimens among the local outcrops, a precursor to his lifelong study of geology and mineralogy. On his return to Boston the budding geologist contributed an article to *Silliman's Journal*, "Notes on the Mineralogy of Nova Scotia," [3] and his trip was reported on in a short article in the *Boston Journal of Philosophy*. [248, pp. 596–598] In the article John W. Webster, the co-editor of the Boston journal, calls attention to the young mineralogist:

> Francis Alger, who had been some months on the peninsula returned with several boxes of different substances, which he had met with while engaged in searching for ores of iron. The examination of these substances has led to the knowledge of a new locality of many highly interesting and beautiful minerals, and Mr. Alger has, with great liberality, allowed us to select from his collection a series of specimens, with permission to offer some account of them. [248, p. 596]

Webster adds the following footnote:

> These have been placed in the cabinet of Harvard College. Mr. A. has had a seal of much beauty cut from one of these crystals.

Francis Alger returned to Nova Scotia in 1827 and 1829 with another of our readers, Charles T. Jackson, to do a more comprehensive geological survey of the peninsula. These two field trips were reported in a long, three-part article in *Silliman's Journal*. [4] The geological duo went back for a third and last time in 1832. During this visit they met and exchanged notes with a Canadian geologist, Abraham Gesner. Upon their return, they published a considerable update of their 1828 report as an article in the *Memoirs of the American Academy of Arts and Sciences* [5] and as a tract, [6]. Figure 6.2 is the bookplate in the Athenæum's copy of the tract noting that it was donated by its authors.

Figure 6.2: Bookplate in the Athenæum's Copy of the Alger/Jackson Tract

There matters stood until 1836 when Gesner published his own *Remarks on the Geology and Mineralogy of Nova Scotia*. [109] Eight fleeting references to the findings of Alger and Jackson are to be found in Gesner's 272-page book, most of them casting aspersions on their findings. In the preface to his *Remarks* Gesner notes that...

> ...the Author has received some information from the remarks of Messrs. Jackson and Alger, of Boston. [109, p. ix]

"received some information," indeed.

While Alger and Jackson thanked some of the local rock hounds[4] in their 1832 tract, they failed to mention Gesner. Paul Lucier conjectures that this was...

> ...probably because they regarded Gesner as a mere amateur, someone with useful information but without scientific or cultural standing beyond Parrsboro. [168, p. 15]

Gesner, like Jackson, was a physician. While making his country doctor rounds he gathered mineralogical specimens throughout the province.[5] Gesner had perhaps

---

[4]Dr. M'Culloch, Count de Bouron, Sir Howard Douglas, Dr. Lincoln, Rev. Mr. King, Messrs. Smith and Brown of Pictou, Mr. Blanchard of Truro, Mr. Carr, Mr. Grant, Mr. Longley, and Dr. McKilay to name a few.

[5]Gesner was the inventor of kerosene and, consequently, is considered by some to be the founder of the modern oil industry. Jackson was an expert witness for the defendant in an action brought by Gesner that turned on whether Albertite, the mined mineral from which Gesner was producing kerosene, was or was not a form of coal. In *Gesner's Dream: The Trials and Triumphs of Early Mining in New Brunswick*, Gwen Martin describes the atmosphere in the courtroom:

felt slighted at not being mentioned by Alger and Jackson in their various reports and consequently had no compliments to return.

After Gesner's book appeared, Alger and Jackson wrote two letters to the Nova Scotia government and its governor at the time, Sir Colin Campbell—one on June 10, 1837, and the other, almost three years later, on February 25, 1840—calling attention to the lack of citation to their work. In the 1840 letter they don't actually use the word 'plagiarize' but they leave little doubt this lack was only due to the respectful tone they wished to set in addressing "your honorable body." The 1840 letter reads in part as follows:

> [A] large portion of his work has been borrowed from them without can-
> did acknowledgement—that their work has served as the model and basis
> of his, that discoveries & observations made by them, either appear as his
> own, or are refered [sic] to others, that in his description of localities, and
> in his statement of facts—although their very language has been adapted,
> following their order step by step in describing the same substances—
> their work is rarely alluded to, except with a view to question or deny
> its accuracy, and that he has copied their Geological Map almost omiting
> [sic] a few important localities, but without adding to it. [242, p. 97]

What seemed to get the goat of the Boston geologists more than anything else was that Gesner's work, "intended for the perusal of the general reader" [109, p. ix] had been read widely and gained him a modicum of notoriety whereas Alger and Jackson's tract had been "gratuitously distributed to scientific men" [139, p. 3] and largely ig-nored by this smaller and presumably more discriminating audience. The whole story is well-told in the first chapter of Paul Lucier's book, *Scientists & Swindlers: Consulting on Coal and Oil in America, 1820–1890* [168].

Bibliographic details about the letters written by Alger and Jackson to the Honor-able House of Assembly are added by Peter von Bitter in his two articles in *Geoscience Canada*, [242] and [243]. Based on his analysis of the handwriting in the draft of the 1840 letter, von Bitter concludes that the letter was originally written by Jackson but toned down a bit by Alger before being posted to Halifax. Figure 6.3 is a clip from the draft of the 1840 letter showing Alger's edits to Jackson's text.[6] The text of the clip with Jackson's original in brackets is as follows:

A key anti-Gesner witness was the American geologist Charles T. Jackson, who loathed Ges-ner for having borrowed heavily from his own reports on Nova Scotia geology. The court case gave Jackson a good shot at revenge and he made the most of it. [170, p. 92]

Jackson testified that Albertite was a form of coal. Gesner lost the case along with a considerable amount of money. In case you're wondering, Albertite is *not* a form of coal.

[6] Thanks to Peter von Bitter, Senior Curator Emeritus of Natural History at the Royal Ontario Museum in Toronto for his kind permission to reprint the draft of the Alger/Jackson letter.

and who have advised them to offer some public statement of facts in defence of their rights [as original discoverers and describers of] and claims as bringing into notice one of the most interesting regions known in Geology. This they [have] prefered [*sic*] not to do though as any public journal, and in fact untill [*sic*] recently they had decided not to trouble the public with the subject in any way, [leaving the dates of the two works] leaving both works to speak for themselves.

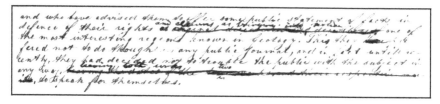

Figure 6.3: Alger's Corrections to Jackson's Draft

Francis Alger would go on to publish a new edition of one of the most widely-read books on mineralogy, Phillips *Elements of Mineralogy* and inherit his father's iron business in South Boston.

6.3   Purposeful Reading about Mineralogy

The Athenæum's 1810 catalog allocated one of its fifteen top-level classifications to Chemistry and Mineralogy placing this class on a par with Theology and Fine Arts.[7] Twenty-two titles are listed under Chemistry and Mineralogy, six of which are the mineralogy titles shown in Table 6.4.

---

[7]See Table A.2 in Appendix A for a list of all the categories in the 1810 catalog along with a count of the number of titles in each.

Author	Short Title
Bergman	*Outlines of Mineralogy*
Brochant	*Traité Élémentaire de Minéralogie*
Brongniart	*Traité Élémentaire de Minéralogie*
Cronstedt	*Essay on Mineralogy*
Kirwan[8]	*Elements of Mineralogy*
Schmeisser	*System of Mineralogy*

Table 6.4: Mineralogy Titles in the Athenæum's 1810 Catalog

Fifty-seven readers generated 114 charges against sixteen mineralogy titles. Figure 6.4 is the cumulative charge plot for mineralogy and Table 6.5 describes the single period of slightly elevated interest. Table 6.2 lists the popular titles in mineralogy and Table 6.5 lists the top readers in mineralogy.

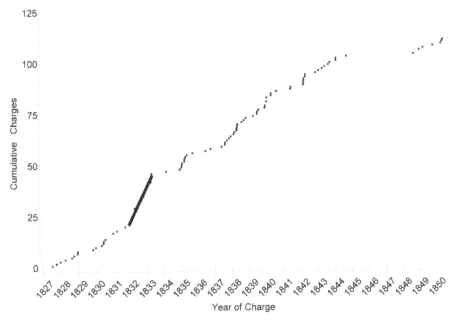

Figure 6.4: Plot of Charges in Mineralogy

---

[8]It was Kirwan's library that ended up at the Salem Athenæum and kindled the interest in mathematics of a young Nathaniel Bowditch.

Period	Start	End	Readers	Charges	Interest
	Jan. 1827	Dec. 1850	57	114	13
A	Dec. 1831	Mar. 1833	9	26	49

Table 6.5: Periods of Elevated Interest in Mineralogy

Author	Short Title	Readers	Charges
Cleaveland	*Treatise on Mineralogy and Geology*	21	34
Dana	*Mineralogy and Geology of Boston*	13	14
Jacob	*Precious Metals*	11	13
Phillips	*Elementary Introduction to Mineralogy*	10	11
Bakewell	*Introduction to Mineralogy*	7	8
Brooke	*Familiar Introduction to Crystallography*	6	8
Welsh	*Familiar Lessons in Mineralogy*	5	7
Varley	*Conversations on Mineralogy*	5	5
Jameson	*System of Mineralogy*	4	4
Mohs	*Treatise on Mineralogy*	4	4

Table 6.6: Popular Titles in Mineralogy

Reader	Charges	Reader	Charges
Parkman, Francis	8	Greene, Nathaniel	4
Codman, Henry	7	Blake, Edward	3
Tuckerman, Edward	7	Alger, Francis	3
Alger, Cyrus	5	Dexter, Franklin	3
Jenks, William	5	Whitney, Moses, Jr.	3
Hall, Joseph	4	Gay, Martin	3
Parsons, William	4		

Table 6.7: Top Readers in Mineralogy

6.3.1   Francis Alger and Phillips's *Mineralogy*

Francis Alger's updated and annotated edition of Phillips's *Elementary Treatise on Mineralogy* [193] was published in Boston in 1844. This edition of Phillips's book was the fifth, "containing the latest discoveries in American and Foreign Mineralogy; with Numerous Additions to the Introduction" according to its title.

Alger borrowed 445 books of all sorts between January 30, 1843, and February 11, 1851, but his reading in experimental philosophy was concentrated in the first fifteen months, January 30, 1843, to April 22, 1844, when he was working on the fifth edition. Figure 6.5 is Alger's page in Volume II of the books borrowed registers during this period and Table 6.8 lists the journal charges on the page.[9]

In the preface to his edition of Phillips, Alger doesn't say when he started the project to update the text but he does say that he abandoned work on it when a competing edition by John Webster at Harvard was announced by Webster's publisher. However, when he learned…

> …that the publisher who had announced Dr. Webster's work as in press, had suspended its publication; and, eventually, that there was little prospect of his proceeding with it. [193, p. i]

Alger decided to take his project off the shelf and carry on.

Further on in the preface, Alger enumerates the extensive changes he made to Allan's fourth edition. He starts by noting:

> The matter now introduced rather exceeds three hundred pages. It comprises over one hundred additional figures in the Introduction, and the Descriptive part; with new species, foreign and American, brought into notice since the date of the last edition; and the addition of many foreign, as well as all the important localities. [193, p. iv]

---

[9]Note in passing that Volume 21 of the 3^d of 1842 of the Philosophical Magazine is marked as "Lost." That there is no apparent trace of a date underneath the word 'Lost' might indicate that there was a means of erasing entries in the registers.

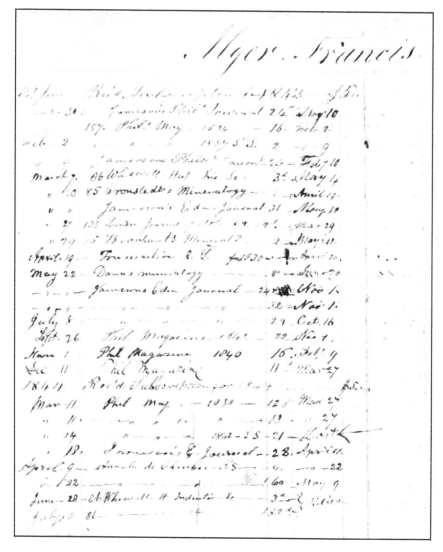

Figure 6.5: Francis Alger's Register Page While Editing Phillips's *Mineralogy*

Charge Date	Volume(s)
Edinburgh Journal	
Jan. 30, 1843	v.24
Feb. 9, 1843	v.29
Mar. 20, 1843	v.31
May 22, 1843	v.24, 29, 32
Mar. 18, 1844	v.28
Philosophical Magazine	
Jan. 30, 1843	v.16
Feb. 2, 1843	v.2 (3s)
Sept. 26, 1843	v.16
Dec. 11, 1843	v.11
Mar. 11, 1844	v.12, 13
Mar. 14, 1844	v.21 (3s)
Annales de Chimie	
Apr. 9, 1844	v.9
Apr. 22, 1844	v.60

Table 6.8: Francis Alger's Journal Charges While Editing Phillips's *Mineralogy*

Francis Alger's father, Cyrus Alger, was also an avid reader in experimental philosophy at the Athenæum. The older Alger was a successful Boston iron founder and undoubtedly had his own interests in geology, metallurgy, and mineralogy so there is no telling if the books and journals he borrowed were for his own use or on behalf of his son, particularly during the period when his son was updating Phillips's *Mineralogy*. Cyrus Alger's periodical charges, including a conjecture as to what article may have interested him, are listed in Table 6.9.

Charge Date	Journal	Possible Article of Interest
Apr. 10, 1843	Edinburgh Trans.	Haidinger, "Mineralogical Account of the Ores of Manganese"
Apr. 20, 1843	Brewster's	Potter, "On the Specific Heats of certain of the Metals"
Apr. 29, 1843	Annales de Chimie	(probably for his son)
Sep. 19, 1843	British Assoc. Rpts.	Hodgkinson, "On the relative Strength and other mechanical Properties of Cast Iron obtained by Hot and Cold Blast"
Sep. 19, 1843	British Assoc. Rpts.	Mallet, "Action of Sea and River Water upon Cast and Wrought Iron", Part 1
Oct. 9, 1843	British Assoc. Rpts.	Mallet, "Action of Sea and River Water upon Cast and Wrought Iron", Part 2
Jan. 9, 1844	Phil. Trans.	Daniell, "On a New Register-Pyrometer, for Measuring the Expansions of Solids, and Determining the Higher Degrees of Temperature upon the Common Thermometric Scale"
Jan. 22, 1844	Phil. Trans.	Hodgkinson, "Experimental Researches on the Strength of Pilars of Cast Iron, and Other Materials"
Jan. 30, 1844	Phil. Mag	Proceedings of the Geological Society

Table 6.9: Cyrus Alger's Periodical Charges in Volume II

There are three sets of footnotes in the fifth edition of Phillips's *Mineralogy*: first, those of the original author, Phillips; second, those of the editor of the fourth edition, Allan; and third, those of the editor of the fifth edition, Alger. Allan's footnotes are marked [E. ED.] and Alger's are marked [AM. ED.]. Of the eighty-three footnotes by Alger, twenty-three include a citation to a specific periodical article and of these five are to an article in one of the journal volumes he borrowed from the Athenæum.

As Alger noted in his preface, additional footnotes were by no means his only contribution to the fifth edition. According to an anonymous review of the book in *The North American Review*:

> Mr. Alger has supplied many deficiencies, corrected many errors, given new measurements of crystals, new chemical analyses, and made most valuable additions to the introductory chapters. Especially has he enhanced the value and utility of this part of the work by the revision of the whole subject of crystallography, and the intelligible manner in which he has explained the new doctrines of Pseudomorphism, Isomorphism, and Dimorphism, of which so little was known when Mr. Phillips prepared his last edition, and which were entirely overlooked by Mr. Allan. [85, p. 241]

6.4 Purposeful Reading about Anthracite Iron

Only one title concerning the manufacturing of iron with anthracite coal appears in the first four volumes of the Athenæum's books borrowed register. The book, *Notes on the Use of Anthracite in the Manufacture of Iron* [141] by Walter Johnson, was charged twice. The first time was on December 16, 1841, by Stephen Higginson, one of the partners in Perkins' Vermont mining misadventure. The second time was on March 28, 1842, by Rufus Wyman. But books weren't the only way to find out about anthracite iron.

On December 10, 1831, William Lyman borrowed Volume 2 of the 1831 edition of the *Franklin Journal*.[10] Table 6.10 lists three articles in the volume that have been of interest to Lyman at the time.

---

[10] An enumeration of all of Lyman's periodical charges can be found in Appendix F.

Page(s)	Short Title
109–110	Patents (3): Improvements in Metallurgical operations
112	Patent: Making Coke from Anthracite Coal
123	Proposed plan for Smelting Iron Ore with Anthracite Coal

Table 6.10: Articles on Making Anthracite Iron in the 1831 *Franklin Journal*

The inaugural issue of the *Franklin Journal* included a list of premiums that would be awarded at the third annual exhibition of the Franklin Institute one of which would be awarded...

> ...[t]o the person who shall have manufactured in Pennsylvania, the greatest quantity of Iron from the ore, using no other fuel but anthracite, during the year ending September 1, 1826. The quantity not to be less than twenty tons.—A Gold Medal.[143, p. 7-8]

There were two additional premiums pertaining to anthracite coal in the list, one for a grate for using anthracite coal at home[11] and the other for using anthracite coal to generate steam. Considerable deposits of what was referred to as stone coal had been discovered in Pennsylvania but the difficulty in igniting it[12] was well-known and proving to be a impediment to its wider use.

In the event, William Lyman would collect a premium for making iron with anthracite coal but not the one offered by the *Franklin Journal*. In his survey, Johnson recalls:

> The Pottsville furnace is the same with which Mr. William Lyman made his experiments....The continuous blast of three months, required by the conditions under which he received this furnace property, was completed in January, 1840. [141, p. 38]

This was also the condition of the prize that Lyman won in 1839.

6.4.1   Lycoming Coal, Farrandsville Nails and Pottsville Iron

In 1828 the directors of the Lycoming Coal Company in Lycoming County, Pennsylvania, decided to put their company up for sale. The prospectus for the sale begins in a business-like manner:

---

[11]"Tastefulness of design, although not a primary objective, will be considered..." This premium was won by Mr. Joseph Page of Philadelphia: "his stove is simple in its construction, and contains all the required conveniences:—The silver metal." [142, p. 264]. The other two premiums for the use of anthracite coal were unclaimed.

[12]...hence stone coal.

> The object of the following brief exposition, is to furnish such facts respecting the coal trade generally, and such a description of the Company's Property, and of the various advantages which it combines, as will enable those who take an interest in such matters, to form a tolerably correct estimate of its value. [253, p. 3]

Anne Knowles picks up the story of the sale in *Mastering Iron*:

> In the spring of 1832, four "men of Boston," as they described themselves, took the bait and purchased the Lycoming Coal Company. [153, p. 119]

Note the phrase "took the bait." The four men from Boston were Thomas H. Perkins, Edward Dwight, Patrick Tracy Jackson, and George W. Lyman, all save Perkins readers in experimental philosophy at the Boston Athenæum.

The railroads that these Boston industrialists were building back East ran on coal and there just wasn't enough to be had from New England deposits to power their trains. In addition to coal, railroads required vast amounts of iron—for tracks, for boilers, for engines, and for rolling stock—and iron furnaces were also ravenous consumers of the stuff. The 1828 prospectus coyly notes:

> In the case of the Lycoming Mines, the value is greatly enhanced by the juxtaposition of Bituminous Coal, Rich Iron Ore, and Lime Stone easily obtained. [253, p. 3]

The Lycoming field was even said to contain the clay that would be needed to make the fire-bricks that would be used to build the furnaces.

The Pennsylvania business plan of the men of Boston was to own the coal mines that supplied coal, iron, and limestone to the iron furnaces that they would build with the clay in their field which would make iron for the railroads they were financing which would run on coal from the mines that they owned to transport coal and iron from their mines and furnaces to East coast markets. With all this synergy, what could possibly go wrong?

At about the same time that Perkins and friends bought the Lycoming Coal Company, William Lyman[13] and another of our readers, Robert B. Forbes, were setting up a factory to make nails from the iron that Perkins' furnace would produce. Whether it was because Perkins' iron was never forthcoming or because marketplace for nail became too competitive, a few years later Lyman built his own furnace in Pottsville, Pennsylvania; a furnace to make iron using anthracite coal.

---

[13] Henceforth, standing alone 'Lyman' will refer to William Lyman, not to George W. Lyman.

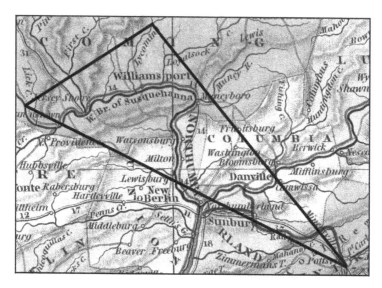

Figure 6.6: Map of Boston Ventures in the Pennsylvania Coal Fields

Figure 6.6 shows the geography of all these business ventures on an extract from Henry S. Tanner's 1836 map of Pennsylvania.[14] At the top of the map is the location of the Lycoming Company's coal field, the source of the coal and iron ore intended to feed the furnaces and eventually the railroads. At the left side of the map is location of the town that an agent of the Lycoming Coal Company, William P. Farrand, founded in 1832, Farrandsville.[15] It was in Farrandsville that the four men from Boston would build their furnace and where Lyman and Forbes would build their nail factory. Farrandsville is hard on the west branch of the Susquehana River and in 1835 the town became the terminus of the West Branch of the Pennsylvania Canal; all very handy for transporting coal, iron, and nails to Philadelphia and other East coast markets. In the lower right of the map is Pottsville, where Lyman would build his anthracite steam furnace.[16]

---

[14]The history of the rise and fall of all of these ventures and many others is told in full color in Ann Knowles' *Mastering Iron* [153]. Carol Brundy's biography of Charles Russell Lowell, Jr. [48] contributes additional details, particularly the failure of Charles Russell Lowell, Sr., in managing the Farrandsville furnace and the disastrous impact that the failure of the whole Lycoming venture had on the fortune of his father-in-law, Patrick Tracy Jackson. After returning to Boston, Lowell became an assistant librarian at the Boston Athenæum and starting in 1857 began working on the Athenæum's last printed catalog. [69] The librarian who finally brought this staggering project to fruition, Charles Cutter, noted in his short history of the effort: "Mr. Lowell was busily engaged in removing defects which he alone knew of, when in June 1870 he suddenly died" [69, p. 3399]

[15]To calibrate the map, it is about thirty-five miles from the Lycoming coal fields to Farrandsville. The town must have been too small or its future too tenuous for inclusion on Tanner's map.

[16]The furnace is often referred to as the Pioneer furnace but it only assumed this name in 1840 when it was purchased by Dr. G.G. Palmer.

Figure 6.7 is an advertisement announcing the opening of the Franklin Nail Works store on August 19 1835. Knowles notes that some time after 1836...

> ...William Lyman and his partner Robert Bennet Forbes shut down the Franklin Nail Works, in which they had invested roughly $50,000. The mill never did make nails from Farrandsville iron. [153, p. 135].

This may be because the Farrandsville furnace never produced any suitable iron. According to a survey of iron and coal trade of Pennsylvania report on in the July, 1846, issue of the *Franklin Journal*, the Franklin Nail Works ran twenty nail-making machines that had produced 10,000 kegs of nails. [54, p. 135]

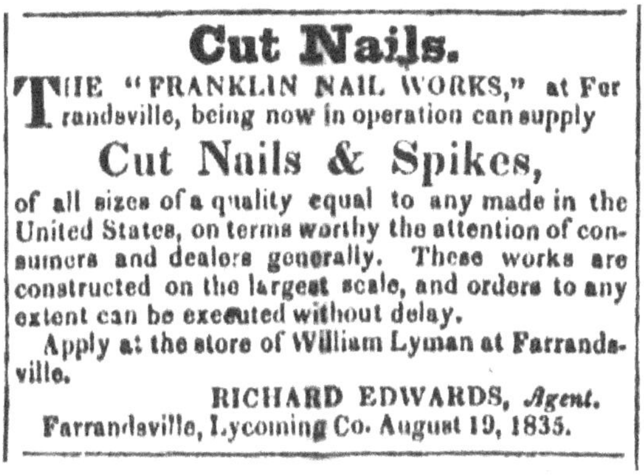

Cut Nails.

THE "FRANKLIN NAIL WORKS," at Farrandsville, being now in operation can supply

Cut Nails & Spikes,

of all sizes of a quality equal to any made in the United States, on terms worthy the attention of consumers and dealers generally. These works are constructed on the largest scale, and orders to any extent can be executed without delay.

Apply at the store of William Lyman at Farrandsville.

RICHARD EDWARDS, *Agent*.

Farrandsville, Lycoming Co. August 19, 1835.

Figure 6.7: Advertisement for the Franklin Nail Works

While all of the men from Boston went back home at a loss,[17] Lyman could at least claim some prize money and a niche in the iron hall of fame. As told by James Lambert and Henry Reinhard in their history of Pennsylvania coal country [156] as well as reported in newspapers of the day:

> The next furnace to use anthracite was the Pioneer, built in 1837 and 1838 at Pottsville, by William Lyman, of Boston, under the auspices of Burd Patterson, and blast was unsuccessfully applied July 10, 1839. Benjamin Perry, who had blown in the coke furnace at Farrandsville, then took charge of it, and blew it in October 19, 1839, with complete success. This furnace was blown by steampower. The blast was heated in ovens at

---

[17]There is a whole chapter in Seaburg and Paterson's biography of Thomas Handasyd Perkins [207] on his Monkton Iron adventure but strangely no mention of either the coal mine and iron furnace in Lycoming county or the lead mine in Northampton.

the base of the furnace with anthracite, to a temperature of 600 degrees, and supplied through three tuyeres at a pressure of 2 to 2% lbs. per square inch. The product was about 28 tons a week of good foundry iron. The furnace continued in blast for some time. A premium of $5,000.00 was paid by Nicholas Biddle and others to Mr. Lyman, as the first person in the United States who had made anthracite pig iron continuously for one hundred days. [156, p. 40]

Perkins' Farrandsville furnace was eventually listed in the National Register of Historic Places in 1991. There was, however, no premium.

In addition to reading about making iron using anthracite coal at the Athenæum, Lyman read about steam engines. Table 6.11 lists Lyman's charges of books in this later subject. Lyman was not interested in the history or science of steam engines, in their use to power railroads, or in their original application, pumping water out of mines. He was interested in using steam engines to blast hot air into an iron furnace that was running on hard-to-ignite stone coal.

Author	Short Title	All Charges	Lyman's Charges
Tredgold	*Steam Engine Investigation of its Principles*	24	9
Farey	*Treatise on the Steam Engine*	23	7
Lardner	*The Steam Engine*	24	1
Partington	*Descriptive Account of the Steam Engine*	4	1
Birkbeck	*Steam-Engine*	1	1
Stuart	*Descriptive History of the Steam Engine*	12	0

Table 6.11: William Lyman's Charges of Books on Steam Engines

The books other than Tredgold's and Farey's are primarily histories of the steam engine. That Lyman preferred Tredgold and Farey[18] supports the view that he was looking for information about the science and technology of steam engines. Tredgold's book is the most mathematical of these two and is centered more on the construction of steam engines than on their use.

Farey sprinkles the history of steam engines throughout his text but uses the history to carry his primary theme which is the advantage of steam engines in

---

[18]John Farey, Jr., was the son of the British geologist to whom discovery of the mathematical construct called the Farey sequence is misattributed. The sequence was first written down by the French mathematician, Charles Haros. See [122].

diverse applications. Of particular interest to Lyman would have been Chapter IV in Farey's book: "On the Manufacture of Iron, and the Application of Cast-iron in the Construction of Steam-engines and Mill Work, and the Application of the Fire-engine to Blow Furnaces and Work Water-Wheels."

In addition to books about steam engines, Lyman logged the twenty-four charges against the periodicals listed in Table 6.12.

Short Title	Charges
American Journal of Science	7
Franklin Journal	7
London Mechanics Magazine	5
Repertory of Patent Inventions	4
Trans. Geological Society of Pennsylvania	1

Table 6.12: William Lyman's Periodical Charges

He also twice charged William Strickland's report to the Society for the Promotion of Internal Improvement in the Commonwealth of Pennsylvania on his 1825 visit to Europe "to collect information of all the valuable improvements in the construction of canals, roads, railways, bridges, steam-engines."

Lyman may have been using the journals as an early social network; that is to find experienced and knowledgable individuals who he could hire to help build and operate his furnace.[19] Furthermore, some of the journals, particularly the *Franklin Journal* and the *Repertory of Patent Inventions*, carried notices of expired and contested patents as well as those newly issued so Lyman may also have been looking for patents to license or to avoid.

In fact, Lyman did license two patents for manufacturing anthracite iron. The first one was patent 7875X issued to Frederick W. Geissenhainer on December 19, 1833, for "iron and steel manufacturing by applying anthracite coal." In the autumn of 1836, Geissenhainer had blown an experimental furnace about ten miles from Pottsville and successfully made a small amount of pig iron using anthracite coal with a hot-air blast. Geissenhainer's patent was purchased from his estate in 1838 by a British iron-man, George Crane. Crane also had a patent for the manufacturing of anthracite iron, US 1024, "improvement in the manufacture of iron" issued on November 29, 1838. Lyman licensed both of these patents from Crane.[20]

---

[19] It was the lack of local expertise not a lack of raw material that doomed the Perkins furnace.

[20] Lyman dealt with Crane's lawyer, Robert Ralston, in negotiating the licensing of Crane's patents. The first puddling furnace in New England was run by Lyman, Ralston & Co. on Boston's mill-dam in the early 1830's. I have not been able to determine if this was perhaps William Lyman and Robert Ralston.

Lyman seems to have set out to make anthracite iron at least as far back in 1831 when he charged the volumes of the *Franklin Journal* noted in Table 6.10. Anthracite coal is advantageous in making iron for technical and economic reasons. It is dense and burns hot so you need less of it to produce a ton of iron as compared to types of coal. Anthracite's primary disadvantage is, as noted above, that it is difficult to ignite. In the operation of an iron furnace, coal, iron ore, and limestone are continuously added to the furnace so ignition is a continuous and ever-present problem. A recently discovered solution to this problem was to heat the air that kept the fire burning before you blasted it into the furnace.

Small furnaces with an abundant supply of local manual labor could operate the bellows manually the same way a blacksmith did. For larger furnaces of the type that Lyman as well as Perkins and friends over in Farrandsville envisioned this was a non-starter. You might be able to run the mills in Lowell with boarding houses full of farm girls but this wasn't going to do for operating the bellows of an industrial-scale iron furnace.

One source of power that had been successfully employed to blow large iron furnaces was a water wheel but the power generated by a water wheel is capped by the power that could be harvested from the flow of the water that drove the wheel. This upper bound constrained how many cubic feet of air per minute you could force into the furnace per minute and consequently the size of the furnace, the type of fuel it consumed, and the amount of iron it produced. There was also the inconvenience of having to locate your furnace on a fast-flowing river. A steam engine, on the other hand, could generate as much power as you wanted and anywhere you wanted it.

W. Ross Yates, wondering about who should get credit as the "principle contributor" to the discovery of making anthracite iron, ends his informative article on the history of the discovery as follows:

> [Making anthracite iron] involved essential contributions from many people—the operators at Vizille, Neilson, Howell, Crane, Thomas, Geissenhainer, Lyman, Perry, Guiteau—these and others put pieces of the puzzle in place. The process was not so much a single invention as it was a new arrangement of old methods, to which some improvements and a few innovations were added. The sparks of genius flickered in a number of people. [261, p. 223]

On February 20, 1836, William Lyman checked out an 1835 volume—either volume 15 or 16, the register doesn't say which—of the *Franklin Journal*. In volume 15 there is a long, five-part article by French mining engineer, Pierre Armand Dufrénoy:

> "To the Board of Directors of Bridges, Public Roads, and Mines, upon the Use of Heated Air in the Iron Works of Scotland and England. By M. Dufrénoy, Engineer of Mines. Paris, 1834. (Translated for this Journal, by S.V. Merrick.)"

It is my conjecture that Lyman's reading of this article contributed in no small way to his decision to pursue making iron using anthracite coal. Lyman traveled back to Pennsylvania in 1837 and along with a local ironmaster, Burd Patterson, started construction of an anthracite furnace that was put into blast on October 29, 1839.

Dufrénoy's treatise on hot air blast is also a thin, tenuous link between Perkins' furnace in Farrandsville and Lyman's furnace in Pottsville.

6.4.2   The Athenæum's Copy of Dufrénoy's *Use of Hot Air*

There is a second book about anthracite iron in the Athenæum's current collection. While it was published in 1834, it doesn't appear in the books borrowed registers because it wasn't added to the collection until 1901. The full title of the book is *On the Use of Hot Air in the Iron Works of England and Scotland. Translated from a report, made to the Director General of Mines in France, by M. Dufrénoy, in 1834* [75]. There is a bookseller label on the front paste-down endpaper that is partially covered by a newspaper clipping. What can be read is the following:

THOMAS MU
Bookseller & S
8, ARGYLL ST
Glasgo

The title page of the book contains the following handwritten inscription:

Thos. Carswell
Muirkirk Iron Works
1837

The back story of this book on the Athenæum's mineralogy shelf might just be the following: Thomas Carswell gained experience in the operation of iron furnaces at the Muirkirk Works in Scotland. On account of his expertise he was lured to Pennsylvania to operate the Farrandsville furnace. The book came to America with Carswell, sat on a shelf in an office in Pennsylvania coal country, and went back to Boston with George W. Lyman. The book was inherited by his son, Arthur Theodore Lyman, who donated it to the Athenæum on June 14, 1901. Figure 6.8 shows the bookplate and the titlepage marginalia in the Athenæum's copy of Dufrénoy's book.

   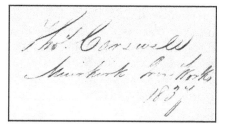

Bookplate   Titlepage Marginalia

Figure 6.8: The Athenæum's Copy of Dufrénoy's *Use of Hot Air*

Knowles contributes a supporting detail to this story:

> Lowell agreed with William Lyman's recommendation that "a Scotchman
> from [the] Muirkirk works would be preferable to a Welchman [*sic*]." The
> Muirkirk Iron Company, founded south of Glasgow in 1787, was one of
> Scotland's largest pig iron producers in the 1830s. [153, p. 128]

That the steam-powered hot-blasts at Farrandsville were made in Glasgow [226, p.
369] might add a little support too.

Chapter 7

Mathematics, Astronomy, and Architecture

7.1   Warren Colburn and the Parallel Axiom

Warren Colburn graduated Harvard in 1820, having "made himself master of the calculus, and read through a considerable part of the great work of Laplace." [91, p. 7]. His thesis, "Calculation of the Orbit of the Comet of 1819," replete as it was with spherical trigonometry and calculus, suggests that one of the works by Laplace that he read was *Méchanique Céleste*, presumably in French as Bowditch's translation wouldn't appear for another fourteen years. Colburn went on to found a school to teach "all branches of pure mathematics, and in their application to navigation, surveying, gauging," to be the superintendent of the Merrimack Manufacturing Company, to invent an improved dynamometer, and, more than all of these, to revolutionize the teaching of mathematics in public schools in America.

   In addition to textbooks, Colburn penned articles on mathematics and its applications for the journals of the day. Table 7.1 lists the articles he published in the short-lived *Boston Journal of Philosophy and the Arts*.

Vol.	Pages	Article Title
I	591–592	"Elements of the Orbit of the Comet of 1823"
II	384–385	"Demonstration of the 47th Proposition of Euclid"
II	474–492	"Demonstration of the Binomial Theorem"
II	201	"Elements of the Comet of 1823"
III	81–91	"A New Theory of Parallel Lines"
III	490–491	"Answer to Professor Strong's "Remarks on Mr. Colburn's Theory of Parallel Lines"

Table 7.1: Colburn's Articles in *The Boston Journal of Philosophy*

The fifth entry in this list, "A New Theory of Parallel Lines," earned him scolding [225] from a professor of mathematics at Hamilton College in Clinton, NY, Theodore Strong.[1] The final entry in the list is Colburn's response to Strong.

The purpose of an experimental essay is to convince you, the reader, that an experiment demonstrates how things really are. If you become convinced, then you can with confidence use the experimental outcome in building railroads, textile mills, hydraulic turbines, and so forth. The cornerstone utility of an experimental essay is that the experimental conditions—for the purpose of this discussion we might think of them as experimental axioms—hold true in the context in which you intend to build.

A mathematical proof is premised on a set of mathematical axioms just as an experimental essay is premised on a set of experimental axioms. A mathematical axiom is a "self-evident proposition, requiring no formal demonstration to prove its truth, but received and assented to as soon as mentioned."[2] In both cases, you either buy into the axioms at the start of the essay or you walk away.

A mathematical proof is in this sense the logical dual of an experimental essay. In an experimental essay, you start with set of experimental conditions and you read through the essay to an experimental outcome in which you become, by the agency of the rules of the experimental essay, morally certain. In a mathematical proof, you start with a set of mathematical axioms and you read through the proof to a mathematical theorem in which you become, by the agency of logical rules of a mathematical proof, morally certain.

The two methods are also similar when it comes to debates about the basic building blocks—the axioms. Both experimentalists and mathematicians acknowledge that it is possible that two axioms which seem as first blush to be different actually amount to

---

[1] Strong and Nathaniel Bowditch were perennial top problem solvers in Robert Adrian's *Mathematical Diary*. Bowditch's son, Henry Ingersoll, was also among the leaders.

[2] See the *Oxford English Dictionary* [90, p. 838] quoting Charles Hutton [137, p. 2].

one and the same thing; that is they are equivalent. In such a situation, it is immaterial which one you pick to work with because you can't build anything more with one than you can with the other.[3]

Faced with two equivalent mathematical axioms, it is largely a matter of fashion which one the mathematical community adopts as the axiom, with the other one becoming a theorem proven using the anointed one. In the nineteenth century, when the gulf between academic and practicing mathematicians was considerably smaller than it is today, an effort was made to adopt as the axiom the one that was self-evident to the greatest number of interested readers. For example, almost everyone, whether they've ever had a course in mathematics or not, would understand and accept as self-evident Euclid's first axiom:

> Things which are equal to the same thing are also equal to one another.

Mathematical axioms that aren't self-evident to anyone but are needed to prove theorems of interest are called postulates.[4] Just like a mathematical axiom, you either grant the truth of a mathematical postulate at the beginning of a mathematical proof or you walk away. And, as with axioms, should it be the case that two postulates are equivalent, preference is often given to the one which is more in harmony with common sense.

But there's the rub. When it comes to mathematical axioms, being 'self-evident' means that a statement is logically true and we all run on the same logic. But when it comes to mathematical postulates, being in accord with one's common sense necessitates that the postulate not run counter to the reader's real world experience and we all come to a mathematical proof with a different portfolio of previous experiences. What is in harmony with one person's common sense may be out of key to another's.

No mathematical postulate illustrates this state of affairs better than the fifth postulate of Euclid's system of geometry, the postulate defining parallel lines. Here's the parallel line postulate that was in tune with Euclid's common sense:

> **Euclid** That, if a straight line falling on two straight lines makes the interior angles on the same side less than two right angles, the two straight lines, if produced indefinitely, meet on that side on which are the angles less than the two right angles.

Read that over again. Does this notion of parallel lines sing sweetly to your common sense? It didn't for a lot of mathematicians who came after Euclid. As a consequence, ever since Euclid wrote down what his common sense said were parallel lines there has been a mathematical cottage industry manufacturing alternative parallel line postulates.[5] And over those same centuries, there has been a competing cottage industry

---

[3] ...or by using them both at the same time, for that matter.

[4] An analogous experimental postulate might be that physical constants really are constant.

[5] See the Wikipedia page for the parallel postulate for a long if not exhaustive list.

trying to show that you could build parallel lines using the first four of Euclid's postulates so you didn't need the fifth one at all.

Here are three parallel line postulates offered up as alternatives to Euclid's by mathematicians of yore:[6]

Aristotle Two convergent straight lines intersect and it is impossible for
two convergent straight lines to diverge in the direction in which
they converge.

Proclus If a line intersects one of two parallel lines, both of which are
coplanar with the original line, then it must intersect the other also.

Playfair Through any point in the plane, there is at most one straight line
parallel to a given straight line.

Anybody can come up with their own parallel postulate but there is one rule in the
game. When somebody proposes a new way of saying what it means for two lines to
be parallel—one that fits their own common sense and one they believe is closer to
the common sense of more people than all the others—they are obliged to prove the
equivalence of their new version with one of the old ones.[7]

Colburn was one such person. In his article "New Theory of Parallel Lines" he set
forth his own definition of parallel lines and, playing by the rule of the game, proved
that his postulate was equivalent to an existing parallel line postulate. Or so he thought.
Strong disagreed. He claimed that Colburn's equivalence proof was faulty and therefore Colburn's new postulate had not been shown to be equivalent to all of the others and therefore could not be admitted into the pantheon of parallel line postulates.
Whether or not Colburn's postulate agreed with Strong's common sense notion of
parallel lines we'll never know. Strong's objection was not with Colburn's parallel line
postulate itself but rather with Colburn's equivalence proof.

Here's Colburn's proposed parallel line postulate:

Colburn Two straight lines so situated as to be equidistant throughout
their whole extent, are called parallel lines.

In other words, two lines are parallel if they are always the same distance apart. Today
this is called the equidistance postulate for parallel lines and has, in fact, been proven
to be equivalent to all the others.

Strong's problem with Colburn's equivalence proof was that in his view Colburn
hadn't proven that two straight lines could be "so situated" as Colburn claimed they

---

[6]A version of the parallel postulate proposed by the Jesuit mathematician Giovanni Girolamo Saccheri
was simply "Rectangles exist." But there are a number of prior axioms and postulates that are needed to
understand why this has anything to do with parallel lines.

[7]This is only true if they want to go on living in the world of Euclidean geometry. See non-Euclidean
geometry for worlds where parallel postulates play that are not equivalent to Euclid's.

could be. In particular, Strong held out the possibility that there might exist a curve that was equidistant from a straight line that was not itself a straight line. Here's the nub of Strong's objection:

> Let AB be a straight line, and suppose CD is another line which is such that the perpendiculars drawn from CD to AB shall always be equal to each other. Now I do not perceive that Mr. Colburn has any where shown that CD must, according to this construction, be a straight line. [225, p. 371]

In his article responding to Strong, Colburn includes Figure 7.1 which is a draftsman's drawing of how his common sense sees parallel lines.

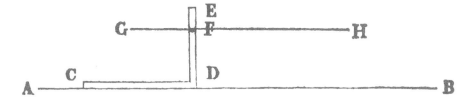

Figure 7.1: Colburn's Definition of Parallel Lines

Colburn introduces the clarification of his constructive proof solicited by Strong as follows:

> Instead of the abstract form I shall give a mechanical description which will not take any from its generality. This mode is often employed by mathematicians in defining curved lines, as circles, ellipses, parabolas, &c.
>
> Let AB be a straight line, and let a *common joiner's square*[8] be supposed to be applied to this line, so that one edge of one leg CD may coincide with it, and the other leg DE will of course be perpendicular to the line.

Warren Colburn was a mechanic before he went to Harvard and he was a mechanic after he left Harvard. To try to convince Strong that two lines could be so situated he is going to use mechanical means, in particular a joiner's square. In mathematics this kind of proof is called a constructive proof for obvious reasons; you actually build one of what you're talking about. Although, it must be said, not typically with mathematical instruments.

Colburn describes the experimental conditions (the two lines and the joiner's square), the experiment (sliding the joiner's square along one of the lines), and the

---

[8]Emphasis the current author's.

experimental outcome (the second line is always the same distance from the first). Colburn even slyly notes that his constructive method of doing mathematics should not be unfamiliar to Strong since "[t]his mode is often employed by mathematicians." To be sure, Colburn's experiment can only be conducted in your head since it requires two lines of infinite length as part of its experimental apparatus and, if that isn't enough, an infinite amount of time in which to conduct the experiment.[9]

It is doubtful that Boyle would have regarded Colburn's article as an experimental essay but he quite possibly might have been willing to grant that Colburn's proof is a rare example of experimental philosophy applied to pure mathematics.[10]

7.2   Purposeful Reading about Mathematics

Reading in mathematics was much more diffuse than reading in the other subject-matter domains. First, there are a comparatively large number of mathematics titles and no handful of these attracted the lion's share of the interest. This may simply be due to the fact that mathematics as a subject matter is itself broader and more diffuse than, say, hydraulics or botany. Second, interest was spread evenly over time so that there is no period during which there was a notably elevated or depressed interest in the topic.

Eight-two readers generated 227 charges against eighty-six mathematics titles. Figure 7.2 is the cumulative charge plot of mathematics titles.[11] Table 7.2 lists the readers in mathematics with more than five charges and Table 7.3 lists the ten mathematics titles borrowed more than five times. Obviously, the other seventy-six titles had fewer than five charges each.

---

[9]...or a joiner's square that slides infinitely quickly.

[10]Just in case you thought that parallel lines were old news, the premiere history of science journal, *Isis*, carried a rave review of a book on parallel lines published in 2014; viz., *La Théorie des lignes Parallèles de Johann Heinrich Lambert*, edited by A. Papadopoulus and G. Théret.

[11]Since the topic is mathematics at the moment, the curve $35.5x^{2/7} - 215$ fits the cumulative charge curve rather nicely.

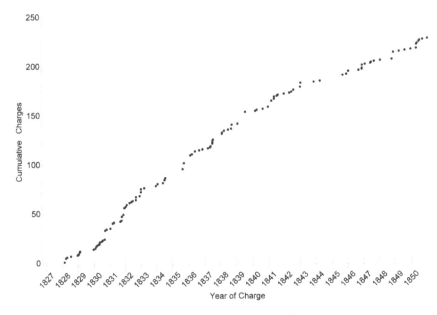

Figure 7.2: Plot of Charges in Mathematics

Reader	Charges	Reader	Charges
Sherwin, Thomas	26	Shurtleff, Samuel	7
Heard, John, Jr.	13	Cooke, Josiah, Jr.	6
Miles, Solomon	10	Bates, George	6
Guardenier, John	8	Ripley, George	6
Young, Edward J.	7	Emerson, George B.	6
Parris, Alexander	7	Capen, Nahum	6
Lawrence, William	7		

Table 7.2: Top Readers in Mathematics

Author	Short Title	Readers	Charges
Priestley	*Familiar Introduction to Perspective*	8	11
Lacroix	*Treatise on the Differential and Integral Calculus*	6	11
Euler	*Letters to a German Princess*	7	9
Hutton	*Mathematical Tables*	4	7
Wright	*Cambridge Problems*	4	6
Bland	*Geometry Problems*	4	6
Simson	*Elements of Euclid*	6	6
Barlow	*Investigation of the Theory of Numbers*	5	6
Euler	*Elements of Algebra*	6	6
Taylor	*Tables of Logarithms*	6	6
Malton	*Compleat Treatise on Perspective*	5	5
Montucla	*History of Mathematics*	1	5
Bourdon	*Elements of Algebra*	3	5

Table 7.3: Popular Titles in Mathematics

Five of our readers were working mathematicians and in addition authored mathematics books of one sort or another: Nathaniel Bowditch, Warren Colburn, Solomon Miles, Thomas Sherwin, and David Bates Tower. Two other readers, Samuel Goodrich and Jacob Abbott, wrote arithmetics. We'll meet Goodwin and Abbott in Chapter 9.

The book by which Nathaniel Bowditch is most widely known today and, it must be added, a book that is still in print today, is his *New American Practical Navigator*.[12] As of October 2016 the book was ranked #61 in Amazon's sailing category. In more rarefied scholarly circles, Bowditch is better known for his four-volume translation of and commentary on Laplace's *Mécanique Céleste*. Bowditch was offered the Hollis Professorship in Mathematics and Natural Philosophy at Harvard in 1806 but he declined.

The books by Colburn and Tower were mathematics textbooks[13] for use in primary and secondary schools as was one of the books published by Sherwin. David

---

[12]See [29] and [30] for the Athenæum's two first edition holdings of what is known widely as simply *Bowditch* as in: "And let me in this place movingly admonish you, ye ship-owners of Nantucket! Beware of enlisting in your vigilant fisheries any lad with lean brow and hollow eye; given to unseasonable meditativeness; and who offers to ship with the Phaedon instead of Bowditch in his head." *Moby Dick,* Chapter 35

[13]The first mathematics textbook written in English by a native American was Isaac Greenwood's *Arithmetick [sic] Vulgar and Decimal: with the Application Thereof to a Variety of Cases in Trade and Commerce*

Bates Tower was a principal at Boston's Eliot Grammar School[14] and at Pennsylvania's Institution for the Instruction of the Blind. In addition to two mathematics textbooks, *Intellectual Algebra* [231] and *Lessons in Oral and Written Arithmetic* [232], Tower wrote a number of grammar and spelling textbooks among which his *First Lessons in Language; or Elements of English Grammar* got a thumbs up review in *The North American Review*:

> Our schools suffer no imposition so egregious as in the cumbrous grammatical text-books in common use, which serve no earthly purpose except to overtask the verbal memory, and to obfuscate the mental perception of the pupil. The Grammar now before us is an honorable exception. [87, p.259].

Tower's co-author on a number of his non-mathematics books, Benjamin F. Tweed, Esq., was a Professor of Rhetoric and Belles-Lettres at Tufts College.

Warren Colburn, George B. Emerson, Solomon Miles, and Thomas Sherwin all graduated in mathematics from Harvard. Table 7.4 lists the year of their graduation as well as the titles of their theses. Emerson was a tutor in mathematics at Harvard and contributed to Farrar's translation efforts but mathematics was not an integral part of his life thereafter.[15]

Reader	Class	Thesis Title
Thomas Sherwin	1825	"Projection of a Solar Eclipse for September, 1838"
George B. Emerson	1817	"Fluxional Solutions of Problems in Harmonicks"
Solomon Miles	1819	"View of the Seat of Ebenezer Crafts, Esquire, in Roxbury"
Warren Colburn	1820	"Calculation of the Orbit of the Comet of 1819"

Table 7.4: Harvard Theses of Athenæum Mathematicians

After graduating in mathematics from Harvard, Emerson, Miles, and Sherwin all started their professional careers as tutors in mathematics at their alma mater. And

---

published by S. Kneeland in Boston in 1727. [12, p.55] Green was the first Hollis Professor of Mathematics and Natural Philosophy at Harvard College.

[14]...established in 1676 and still going strong today.

[15]See [247] and [132].

all of them went on to become headmaster of Boston's English High School. Table 7.5 lists the tenure of each in that position.

Reader	Tenure
George B. Emerson	1821–1823
Solomon Miles	1823–1837
Thomas Sherwin	1837–1869

Table 7.5: Headmasters at Boston English

Colburn and Emerson both borrowed mathematics books from the Harvard College Library during their undergraduate years. The mathematics books they borrowed are listed in Table 7.6.

Author	Short Title	Charge Date
Emerson		
Smith	*Harmonics*	Sept. 4, 1816
Emerson	*Fluxions*	Sept. 13, 1816
Emerson	*Introduction to Mathematics*	Nov. 9, 1816
Bobson	*Mechanical Philosophy*	Nov. 22, 1816
Laplace	*Systétem du Mondé*	Nov. 22, 1816
	Philosophical Trans. v.74, 1747, 1792, 1793, 1800	various
Bailly	*Histoire de l'Astronomie* Tom. IV	Mar. 18, 1817
Legendre	*Théorie des nombres*	May 9, 1817
Euler	*Letters,* v. 1	June 23, 1817
Euler	*Letters,* v. 2	June 30, 1817
Colburn		
Lacroix	*Élements d'Algébra*	Feb. 20, 1817
Lacroix	*Complément d'Algébra*	Mar. 20, 1818
Lacroix	*Élements d'Algébra*	June 2, 1818
Lacroix	*Calcul Différentiel* Tom. I	June 26, 1818
Lacroix	*Complément d'Algébra*	July 17, 1818

Table 7.6: Undergraduate Charges in Mathematics by Emerson and Colburn

John Heard, Jr., whose interest in mathematics was second only to Thomas Sherwin, was a lawyer and a registrar of the Suffolk County probate court. John Guardenier, another avid reader in mathematics, was a book binder as well as a Proprietor of the Boston Athenæum. As recorded on page 220 of Volume II and shown in Figure 7.3, Guardenier paid for two of his Proprietor shares in kind; i.e., by binding books for the Athenæum. I have not been able to connect the considerable interest in mathematics evidenced by these two readers to other parts of their lives.

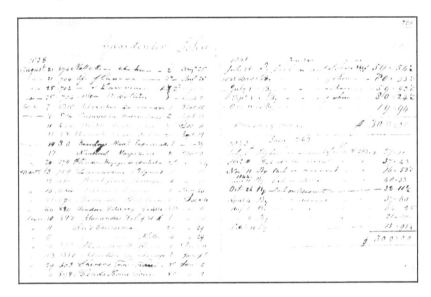

Figure 7.3: John Guardenier's Payment in Kind

7.2.1 Warren Colburn's *Intellectual Arithmetic*

It was Warren Colburn's textbook, *Intellectual Arithmetic, Upon the Inductive Method of Instruction*, that revolutionized the teaching of primary school mathematics in America, by all accounts for the better.[16] An anonymous review appearing sixty-four years after *Intellectual Arithmetic* first appeared effused:

> To commend Colburn's Arithmetic would be like painting the rose. The system he introduced has held its place for sixty years, and educators are not yet ready to depart from its principles. [89, p. 850]

---

[16] See for example [200] and [55].

Historian of mathematics, Florian Cajori, describes the revolution as follows:

> Colburn's *First Lessons* embodied what was then a new idea among us. Instead of introducing the young pupil to the science of numbers, as did old Dilworth, by the question, "What is arithmetic?" and the answer, "Arithmetic is the art or science of computing by numbers, either whole or in fractions," he was initiated into this science by the following simple question: "How many thumbs have you on your right hand? How many on your left? How many on both together?" The idea was to begin with the concrete and known, instead of the abstract and unknown, and then to proceed gradually and by successive steps to subjects more difficult. [50, pp. 106–107]

According to Cajori [50, p. 106] over three and one-half million copies of *First Lessons* were sold both in America and in Great Britain. In his *Bibliography of Mathematical Works Printed in America Through 1850*, Karpinski tabulates forty-two editions in English as well as nine editions in Hawaiian up to 1849 [151, p. 236–39] and adds:

> There were many editions in English up to 1895; the later ones were printed largely in Boston. The work in Hawaiian was printed in Boston as late as 1873.

Colburn's textbook got laudatory reviews when it was published and continues to get complimentary reviews in today's history of mathematics afterglow. Here's David Eugene Smith writing in the *Educational Review* a century on:

> [Colburn's] idea was that a child should think—a revolutionary idea for most teachers of arithmetic in those days....Here, then, was the first great external influence, one based on a mixture of child psychology and common sense, that caused any change in the sluggish course of American arithmetic, and it is one of the few influences that have been exerted on the subject which are really significant. [219, p. 113]

Reader George B. Emerson wrote an introduction to the 1863 edition of Colburn's *First Lessons*. Emerson was not only a friend of Colburn but was a beta-tester for Colburn's revolutionary book. Cajori notes:

> [Colburn] had read Pestalozzi, most probably, while in college. A manuscript copy of his First Lessons was furnished by Colburn to his friend George B. Emerson for use in a school for girls, and the former received valuable suggestions from the latter. The success of the book was almost immediate. No school-book had ever had such sale among us as this. Over three and one-half million copies were used in this country, and it was translated into several European languages. [50, pp. 106–107]

After a brief stint teaching immediately after graduation, in April of 1823 Colburn accepted the position of Superintendent at the Boston Manufacturing Company in Waltham and then, in 1825, he accepted the post of Superintendent at the Lowell Merrimack Manufacturing Company. This is not to say that he lost his enthusiasm for either mathematics or education. He served on the Superintending School Committee in Lowell and gave lectures on a wide range of scientific subjects throughout greater Boston. He was elected a fellow of the American Academy of Arts and Sciences in 1827 and served on the Examining Committee for Mathematics at Harvard.

Colburn crossed paths with two other of our readers of mathematics, Edward Davis and Patrick Tracy Jackson. Davis was a classmate of Colburn and contributed a lengthy "Reminiscences of a Class-Mate" to Edson's memorial to Colburn [92, pp. 7–9]. In the memorial Edson recounts:

> Visiting the families of his pupils, he was introduced to the late Patrick T. Jackson, who, with his quick perception of the qualifications and abilities of men, soon discovered in his new acquaintance the talents and acquirements adapted to a situation which he was then seeking to fill. [92, p. 14]

Table 7.7 lists the first edition of each of Colburn's most popular publications.

Year	Title
1820	*Intellectual Arithmetic, upon the Inductive Method of Instruction*
1821	*An Arithmetic on the Plan of Pestalozzi: with some Improvements*
1822	*First Lessons in Arithmetic on the Plan of Pestalozzi: With some Improvements*
1822	*Arithmetic: Being a Sequel to First Lessons in Arithmetic*
1823	*A Key Containing Answers to the Examples in the Sequel to First Lessons in Arithmetic*
1825	*An Introduction to Algebra upon the Inductive Method of Instruction*

Table 7.7: Mathematics Textbooks by Warren Colburn

7.2.2   Miles' and Sherwin's *Mathematical Tables*

In his address to the 1834 meeting of the American Institute of Instruction on the teaching of mathematics in Boston's English High School and, in particular, the use of what he called the Cambridge course,[17] Thomas Sherwin, at the time the sub-master of Boston English, observed:

> The improvement arising from the introduction of the course at Harvard University was great; it contributed much to the interests of the Institution, and did honor to the gentleman by whose labors and talents it was effected. It is, therefore, with regret, that I see some of the best works of the course rejected, and others of less merit substituted in their place. Whether this change be owing to the opinion of those who have the direction of the studies, that the works substituted are really preferable to those displaced, or that the latter are too difficult for the students, I have no means of judging. But in regard to the former supposition, others may honestly differ from them in opinion; and in regard to the latter, I will merely say, that if boys thirteen years old can learn these books, there seems to be no reason why the students of Harvard College should be inadequate to the task. [211, p. 165]

The Cambridge course referred to by Sherwin included works by Colburn as well as translations of Legendre's *Geometry*, Lacroix's *Algebra*, Euler's *Algebra*, and Lacroix's *Arithmetic*. Some of these translations were done by John Farrar, professor of mathematics at Harvard, with the assistance, in some cases, of reader George B. Emerson.

Boston English was founded in 1821 to endow its graduates with "a foundation for eminence in his profession, whether mercantile or mechanical...a different education from any which our public schools can now furnish." [201, p. 2] Sherwin's veiled reference to "other public schools" was undoubtedly to Boston Latin which at the time was devoted almost exclusively to fitting students for Harvard. While Sherwin's observation is anecdotal, if we look to the early nineteenth century curriculums of Boston English and Boston Latin it is not difficult to understand what he was talking about.

Starting in 1803 admission to Harvard required only arithmetic to the rule of three in mathematics.[18]   As Boston Latin's curriculum was driven by Harvard's admission requirements, it is not surprising that the school's curriculum was light on mathematics.

---

[17]See [79] for an extensive description and review of the Cambridge course in mathematics by Benjamin Silliman.

[18]The rule of three is a rote method to solve problems of the following sort: "If Sally reads three books in seven days how long will it take her to read eleven books?" The pedestrian mathematical landscape must have been littered with such problems in the eighteenth and nineteenth centuries since being able to wield the rule of three was an important milestone in a student's mathematical education. That the problem has vanished today suggests, however, that the problems may have only existed in textbooks.

Nor is it surprising that historians of antebellum mathematics, David Eugene Smith and Jekuthiel Ginsburg, [220] found little of note when they wrote about the mathematical awareness of Harvard's entering freshmen. If they had looked beyond the ivy wall, they would have found the mathematics of navigation, surveying, mensuration, and astronomical calculation and the construction of mathematical instruments as well as the above-mentioned arithmetic of Lacroix, algebra of Euler, geometry of Legendre in the mathematical toolkits of Boston's practitioners. [201, p. 5] The graduates of Boston English were not only better prepared in mathematics than their friends across town at Boston Latin but this may well have still been the case four years later when their friends graduated from Harvard.

Solomon Miles and Thomas Sherwin were lifelong friends. When he was 12 or 13 years old, Sherwin attended a private school in Hollis, New Hampshire, run by Miles. [18, p. 229] In 1830 the two published *Mathematical Tables: Comprising Logarithms of Numbers, Logarithmic Sines, Tangents, and Secants, Natural Sines, Meridional Parts, Difference of Latitude and Departure, Astronomical Refractions*. In the Advertisement to their book they document some background of the effort:

> The Tables, comprised in this volume, have been very carefully compared with the best English and French Tables, and they will be found, it is believed, not inferior, in point of correctness, to any similar Tables in use.
>
> Prefixed is a short introduction, explanatory chiefly of the methods of using them.
>
> For information respecting the theory and construction of logarithms, the student is referred to Lacroix's *Algebra* (Cambridge translation,) to Smyth's *Algebra*, and to the more extended English and French treatises. [176, p. 3]

Unlike almost all other forms of literature, the author of a mathematical table is not only permitted to make a verbatim copy of the work of previous authors, he is strongly encouraged to do so. There is, after all, only one sequence of seven digits that comprise the logarithm to seven places of two [19] so the sequence will be the same whether you copy it or recompute it. Copying entailed less handling of the digits than recomputing them and thus introduced fewer errors into the new table, at least before it got to the printer.

Either way the table maker is obliged, also by the conventions of table making, to check his results against existing tables to make absolutely certain that his values are exactly the same as everybody else's. Best practice in this regard was to say precisely which tables were used. When a discrepancy was encountered it was typically but not always in the new value. Should it be the value in the checking table that was in error,

---

[19] 3010300 to the base 10.

the table maker was obliged to report the erroneous value in his publication. More of this below.

Table 7.8 lists the four editions of the Miles and Sherwin mathematical tables. Running up to the first edition, on November 6 and December 6 in 1829 and again on January 27, March 8, and April 14 in 1830, Miles borrowed the fifth (1811) edition of Hutton's *Mathematical Tables* [136]. On November 6, 1829, Miles borrowed the 1812 stereotyped edition of Callet's *Tables Portatives de Logarithmes* [51]. And, on January 16, 1830, he checked out Taylor's *Tables of Logarithms of All Numbers from 1 to 101000* [228]. We thus have some circumstantial evidence that among "the best English and French Tables" that Sherwin and Miles used to vet their tables were the tables of Hutton, Callet, and Taylor.

Edition	Year	Boston Publisher
1st	1830	Carter and Hendee
2nd	1836	James Munroe and Company
3rd	1842	Benjamin B. Mussey
4th	1868	Taggard and Thompson

Table 7.8: Editions of the Miles and Sherwin *Mathematical Tables*

In addition to not citing exactly which tables they used to check their values, Miles and Sherwin didn't say, as was also best practice, which errata to their checking tables they tracked down and applied. Glaisher [112, p. 335] reports that there are twelve errors in the edition of Hutton, eleven in Callet and six in Taylor. Bowditch in using Hutton's tables reports in the Errata section of *American Practical Navigator* [29, p. 590ff.] under the heading "A list of Errors in Hutton's Logarithms." He notes:

> In making the calculations of the preceding work, the following errors ware discovered in Hutton's *Logarithmic Tables*, edition 2d. London, 1794, and edition 3d. London, 1801.

Table III.
Log. of 101009 for 00836 &c. read 00436 &c.

This is the standard idiom for reporting errors in mathematical tables. It fairly obviously means that the digit sequence 00836 in the table entry for the logarithm of 101009 should instead be 00436. Table III in Hutton's book tabulates the logarithms of the natural numbers to twenty decimal places. The entire tabular entry is

00436 00715 66797 88804.

The error that Bowditch found is in the third decimal place and would certainly affect computations using the value. And because the error is in third place of a twenty-place value, it can undoubtedly be charged to the printer and not to Hutton.

A check of the Miles/Sherwin logarithm table shows that all of the entries in Glaisher's list of the errors in Vega's logarithms from which Hutton was computed and all of the errata recorded by Bowditch have been corrected in the 1830 Miles and Sherwin logarithm tables. All this said, it should be recalled that Babbage read his 1827 tables of logarithms against Hutton, Taylor, and Callet, as well as five other logarithm tables and there were still errors lingering in his tables.[20]

As the second edition of the Miles and Sherwin tables was being prepared for the press, Sherwin borrowed Barlow's *Mathematical Tables* on July 18, 1836 and Taylor's tables on March 24, 1838. Just before the third edition, Sherwin again checked out Barlow's tables on February 10, 1841. Table 7.9 lists all of the charges of logarithm tables by Miles and Sherwin.

Charge Date	Reader
Callet's *Tables Portatives de Logarithmes*, 1812	
Nov. 6, 1829	Miles
Hutton's *Mathematical Tables*, 5th ed., 1811	
Nov. 6, 1829	Miles
Jan. 7, 1830	Miles
Mar. 8, 1830	Miles
Apr. 14, 1830	Miles
Taylor's *Table of Logarithms*, 1792	
Jan. 26, 1830	Miles
Mar. 24, 1838	Sherwin
Barlow's *Mathematical Tables*, 1814	
July 18, 1836	Sherwin
Feb. 10, 1841	Sherwin

Table 7.9: Mathematical Tables Borrowed by Miles and Sherwin

---

[20] See "Raymond Clare Archibald and the Provenance of Mathematical Tables" [123] for a discussion of the persistence of errors in logarithm tables.

7.2.2.1   Bowditch, Babbage, and Tables of Logarithms

Tables of logarithms in the seventeenth, eighteenth, and nineteenth centuries had fully as much an impact on scientific and commercial computations as did electronic computers in the twentieth century, perhaps even more so. Both inventions enabled arithmetic operations to be performed 1) more quickly, 2) with fewer errors, and 3) to greater precision than had previously been practically possible.[21] In the case of printed mathematical tables, all of these benefits were conditional on the tables being free of errors.[22]

Computing the logarithm of a number is not difficult; there are many algorithms for doing so. If you are publishing a table of logarithms the difficulty lies in herding a million or more digits through the press without their being changed or rearranged. As was noted above, the error Hutton's table corrected by Bowditch was much more likely to have been introduced by the publisher than by Hutton. Unlike running text in which an upside down 'm' would be noticed and corrected, an upside down '6' masquerading as a '9' won't catch the eye of anybody in the print shop. This source of errors in mathematical tables was no small part of the reason that Charles Babbage created his mechanical computers. Babbage's engines are justifiably heralded for their computational capabilities but they were not noticeably faster than humans. What distinguished them was a little box on the side, a mechanical printer, that output the results of a computation in a form that could be taken directly to the print shop and made into type. No man-in-the-middle—journeyman, typesetter, compositor, or pressman—intervened between the computations and the printed page. After railing against persistence of errors in printed mathematical tables, including an "Erratum of the Erratum of the Errata[23] of Taylor's *Logarithms*," Babbage grumps:

> If proof were wanted to establish incontrovertibly the utter impracticability of precluding numerical errors in works of this nature, we should find it in this succession of error upon error, produced, in spite of the universally acknowledged accuracy and assiduity of the persons at present employed in the construction and management of the Nautical Almanac. It is only by the *mechanical fabrication of tables*[24] that such errors can be rendered impossible. [8, p. 282]

N.B. "mechanical fabrication" not "mechanical computation."

---

[21] These are exactly the properties of a computational technique you want if you're in charge of an artillery battalion in the heat of battle. There is a table of logarithms in the Athenæum's Knox collection [154] by the French mathematician Nicholas Louis de Lacaille that was created for use by French artillery units. Knox was one of Washington's most capable artillery commanders. One can easily imagine this little 16mo volume in Knox's back pocket up on Dorchester Heights.

[22] It's not that moving from printed tables to digital computers did away with tabular errors. The floating point bug in Intel's Pentium processor was due to an incorrect value in a table that was used to initiate multiplication.

[23] For cos . $4°18'3''$, *read* cos . $14°18'3''$, in case you're wondering.

[24] Emphasis Babbage's.

In addition to correcting an error in Hutton's logarithm table, Bowditch took note of errors in other tables he used while at sea. In the preface to the 1802 edition of his *New American Practical Navigator* Bowditch comments:

> The author had once flattered himself, that the tables of Maskelyne[25] which did not depend on observations would be absolutely correct; but in the course of his calculations he has accidentally discovered several errors in two of the more correct works of the kind extant, viz. Taylor's and Hutton's Logarithms,[26] notwithstanding the great care taken by those able mathematicians in examining and correcting them: he therefore does not absolutely assert that these tables are entirely correct, but feels conscious that no pains have been spared to make them so. Any one who wishes to examine these calculations may do it by the formulas used for this purpose, which will now be given with some additional remarks. [29, p. vi]

Bowditch read his tables against two of the same tables of logarithms that I conjecture were used by Sherwin and Miles, namely, Taylor and Hutton. These were tables of the values of mathematical functions. Bowditch was far less polite when it came to errors in navigation tables:

> Table V. contains the declination of the sun for the years 1802 to 1819, and marked to the nearest minute.—This table is one of the most important in this collection, because the latitude is generally determined by it: it was therefore a very criminal inattention of Moore, in publishing it so incorrectly in most of the late editions of his work; for, by reckoning the year 1800 as leap year, he had made an error of 23 miles in some of the numbers. This error was the cause of losing two vessels to the northward of Turk's Island, and bringing others into serious difficulties. [29, p. vii]

In this case, the error was entirely the fault of the table maker, John H. Moore, and cannot be charged to the printer. The rule for determining if a year is a leap year is that you divide the year by 4 and if the answer comes out evenly (without a fraction) then the year is a leap year *unless* the year ends in 00 in which case you divide by 400. Moore neglected the 'unless' part and took 1800 to be a leap year.[27]

---

[25] Maskelyne's *Requisite Tables*

[26] Footnote in Bowditch: As it will be useful to those who own these works to have a list of these errors, it is given in the last page of this work.

[27] There is an urban legend that early versions of Microsoft's spreadsheet program, Excel, didn't implement the 'unless' part either in order to be compatible with the day's most popular spreadsheet program, Lotus 1-2-3, which followed Moore.

7.2.2.2   The Judgment of the Peirces

Figure 7.4 is the Advertisement page from a Google scan of a Harvard copy of the Miles and Sherwin mathematical tables.

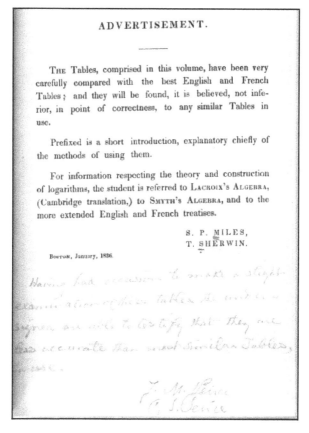

Figure 7.4: Marginalia in Miles & Sherwin's *Mathematical Tables*

At the bottom of the page there is a handwritten note that reads:

> Having had occasion to make a slight examination of these tables the undersigned are able to testify that they are less accurate than most similar tables, in use. /J.M. Peirce/C.S.Peirce

The note is attested to by two Harvard mathematicians, James Mills Peirce and Charles Sanders Peirce.

Paging back to the bookplate page of the scan, one find's the book's call number written at the top—Math 838.42. At the bottom of the bookplate page is a partially obscured stamp that read "Transferred to Cabot Sc." On the last page of the digital file the bar code tag on the outside of the back cover of the book is clearly visible:

Math 838.42
Mathematical tables;
Cabot Science

Consulting Hollis, the online catalog of Harvard's libraries, one finds that there are four copies of the Miles and Sherwin tables in Harvard's collection. Three of them—the 1830 first edition, the 1836 second edition, and the 1842 third edition—were listed as being on the shelf at Cabot Science library. The fourth, another copy of the 1842 third edition, was at Widener. As you have undoubtedly guessed by now, the 1842 third edition on the shelf at Cabot was indeed the copy that had been digitized by Google and contained the handwritten review signed by the Peirces.

There is no telling which of the seven tables in the Miles and Sherwin book the Peirces examined but we can be sure that neither of the Peirces would confuse accuracy and precision. The three largest tables in the book—logarithms of natural numbers, logarithms of trigonometric values, and natural trigonometric values—are all to seven-places. While there are certainly tables of these values to a greater number of decimal places, seven places was by far the most widely-published level of accuracy for logarithm tables.

In fact, exactly what the Peirces were complaining about is a bit of a puzzle. It certainly doesn't take more that a slight examination to determine that the tables are to seven places so this can't be reason for their criticism. The tables after all don't claim to be accurate to more than seven places. On the other hand, it would take more than a slight examination to determine that there were errors in any of the Sherwin/Miles tables.

As the Peirces' marginalia is undated we don't know when they conducted their informal review. We do know that James M. Peirce published a 14-page tract, "Three and Four Place Tables of Logarithmic and Trigonometric Functions"[191] in 1871. In a separate explanation of the tables in his tract, Peirce gives a very short history of logarithms and logarithm tables but he doesn't mention the provenance of the numbers in his tables. It is well within the range of possibility that the Peirces read James' tables against the Miles and Sherwin tables and did indeed find errors. If this were the case, then best practice among table makers would, as noted above, be to report the errors and their corrections in the tract just as Bowditch reported the error he found in Hutton.

The copy of the Miles and Sherwin tables with the Peirce marginalia has been transferred to Harvard's rare book collection in Houghton library.

7.2.3   Sherwin's *Elementary Treatise on Algebra*

On June 3, 1834, Thomas Sherwin borrowed *Memoir of Pestalozzi* [172] by Charles Mayo. Two days later he borrowed *Bland's Algebraical Problems* [17]. Sherwin would go on to publish two algebra textbooks, both of them "on the plan of Pestalozzi." The first, *An Elementary Treatise on Algebra for the use of Students in High Schools and Colleges*[212], was published in 1842 and the second, *The Common School Algebra* [213], three years later. The second title achieved a greater success than the first and was among the texts on which Harvard freshman were examined in the 1850s and 1860s—only to Section XXXVIII, Transformation and simplification of irrational quantities, however.[28]

In preparing these textbooks, Sherwin borrowed texts on algebra from the Boston Athenæum. The dates and titles of these charges are listed in Table 7.10.

Author	Short Title	Charge Date
Bland	*Algebraic problems*	June 10, 1834
Bland	*Algebraic problems*	Aug. 6, 1836
L'huiler	*Algébre*	Aug. 9, 1836
Bourdon	*Algébre*	Sept. 9, 1836
Bourdon	*Algébre*	Aug. 28, 1837
Bridge	*Algebra*	Jan. 28, 1837
Euler	*Algebra*	Apr. 24, 1839

Table 7.10: Algebra Books Borrowed by Thomas Sherwin

Sherwin concludes the Preface to the first edition of *An Elementary Treatise* by acknowledging his debt to other mathematicians of the day:

> The writer is unwilling to close his remarks, without expressing his obligations to others, who have done so much to introduce into our country a natural and rational mode of studying mathematics. Among these none merits greater praise than Colburn; and his works have served as a guide in the composition of several others on the inductive plan. Day, Smyth, Davies, and Peirce[29] deserve also to be mentioned with great respect. [212, p. iv]

---

[28]Sherwin co-authored a report in 1848 regarding the mathematics component of the Harvard entrance examination [214] and was himself on the examination committee.

[29]Jeremiah Day of Yale, William Smyth of Bowdoin, Charles Davies of West Point, and Benjamin Peirce of Harvard.

Sherwin's debt to Colburn is particularly strong. *An Elementary Treatise on Algebra* could reasonably be regarded as the secondary school follow-on to Colburn's elementary school *Introduction to Algebra upon the Inductive Method of Instruction*. The tables of contents of the two works are identical for practical purposes.

In *Treatise on Algebra* Sherwin uses the computation of the logarithm of a number as an example of the use of two handy mathematical methods: mean proportionals and mean differentials. He starts with the simplest of all logarithm tables much as the inventor of logarithms, John Napier, did two centuries before:[30]

N	$\log N$
1	0
10	1
100	2
1000	3
10000	4

Table 7.11: A Simple Logarithm Table

To multiply two numbers using this table, find them both in the left-hand column, move to the right-hand column, add the two numbers you find there, and then move back to the left-hand column from the sum to find the product of the two you started with. For example, to multiply 10 and 100, move to the right-hand column to find 1 and 2, add these together to get 3, and then move back to the left-hand column from 3 to find the answer, 1000. Now all Sherwin has to do is fill in the gaps between the entries to get a more useful table. In this case however, linear interpolation will not do; the logarithm of 55, halfway between 10 and 100, is about 1.74 not 1.5, for example.

The gaps in the left-hand column of the simple logarithm table are filled in with an intermediate value using the mean proportional. The mean proportional between two numbers a and b is a computed value between the two like an average but different. In particular, it is the number c such that a is to c as c is to b;[31] that is,

$$\frac{a}{c} = \frac{c}{b}$$

or, what is the same thing,

$$c = \sqrt{a \times b}.$$

---

[30] This is not strictly true as Napier's original logarithms were not to the base ten.
[31] See Rule of Three on page 156.

The gaps in the right-hand column of the table are filled in with a different intermediate value, the mean differential. The mean differential between two numbers a and b is also a computed value between the two like the mean proportional. It is the number c such that

$$c = \sqrt{\frac{b}{a}}.$$

Quoting from page 261 of Sherwin's *Treatise on Algebra*:

Thus, the mean proportional between 1 and 10 = $\sqrt{1.10}$ = 3.162277; and the mean differential between 0 and 1 = $\frac{1-0}{2}$ = $\frac{1}{2}$ = .5.

Thus, applying mean proportional and mean differential to create a new row in the above starter logarithm table, Table 7.11, we get:

N	$\log N$
1	0
3.162277	0.5
10	1
100	2
1000	3
10000	4

Table 7.12: An Improved Logarithm Table

This method of building a table of logarithms produces printed pages that have an unfamiliar look. Normally, we expect to see the numbers whose logarithms are given in the table[32] increase in a regular, step-wise manner as, for example, Table 7.13.

---

[32] These are called the *arguments* of the table.

N	$\log N$
1.000000000	0
1.232846739	0.0909090909
1.519911083	0.1818181818
1.873817423	0.2727272727
2.310129700	0.3636363636
2.848035868	0.4545454545
3.511191734	0.5454545455
4.328761281	0.6363636364
5.336699231	0.7272727273
6.579332247	0.8181818182
8.111308308	0.9090909091
10.00000000	1.0000000000

Table 7.14: Sherwin's Unusual Logarithm Table

N	$\log N$
1.000000	0
2.000000	0.3010300
3.000000	0.4771213
4.000000	0.6020600
5.000000	0.6989700
6.000000	0.7781513
7.000000	0.8450980
8.000000	0.9030900
9.000000	0.9542425
10.00000	1.0000000

Table 7.13: A Familiar Logarithm Table

In a logarithm table that Sherwin's teaching algorithm produces, Table 7.14 for example, the logarithm values increase in a stepwise manner while the values whose logarithms these values are fall where they must. Sherwin's table is unusual but not unusable. One can just as easily interpolate between 2.310129700 and 3.511191734 to find the logarithm of 2 as one can between 2 and 3 to find the logarithm of 2.5.[33]

Sherwin concludes his section on computing logarithm tables by cautioning:

> This process which we have given, is designed to show the learner the possibility of constructing logarithms, rather than as a mode which can conveniently be reduced to practice.
>
> The methods by which logarithms are actually calculated, are in general very different from that given above, and are too complicated to be introduced into an elementary treatise. [212, p. 262]

Sherwin tipped his hat to William Smyth in the preface to *Treatise on Algebra* and also referenced Smyth's *Algebra* in the Advertisement of the Miles and Sherwin mathematical tables:

> For information respecting the theory and construction of logarithms, the student is referred to Lacroix's Algebra (Cambridge translation), to Smyth's Algebra, and to the more extended English and French treatises. [176, p. iii]

The methods of Lacroix and Smyth for computing a logarithm are the same and both of them are based on Lagrange's *Traité de la résolution des équations numériques de tous les degrés* [155].[34] Lagrange is widely respected for the clarity of his mathematical expositions but at the same time is criticized by pure mathematicians for a lack of rigor and formalism.

A Lagrangian mathematical experimental essay begins with elementary example of the theorem he wishes to demonstrate. The introductory example is followed by a sequence of increasingly involved examples that ends with a general statement of what is going on that summarizes all of the examples not unlike Tredgold's Rules. The general statement is the theorem that Lagrange set out to demonstrate[35] and the examples are the experiments that convince the reader of its truth.

The advantage of Lagrange's way of doing mathematics from a practitioner's point of view is that the reader comes to a working understanding of why and how a theorem works. And, as is *de rigueur* for a Boylean experimental essay, if you aren't morally certain about the truth of one of his theorems, you can easily walk through an example or two of your own to convince yourself. And just as with experimental essays such as those by Tredgold and Evans, you can use Lagrange's examples as step-by-step guides in applying his theorems to your own problems.

---

[33] …well, almost as easily. Either way straight linear interpolation between tabular values still won't do and this is why logarithm tables include little sub-tables called difference tables.

[34] The texts of Lacroix and Lagrange are both in the Athenæum's collection. Smyth's is not.

[35] …mathematical formalists refer to Lagrange's discourses as demonstrations not proofs.

In the quest to achieve the greatest generality, the formal proofs of academic mathematicians obscure the inner working of a theorem behind a complex rhetorical scaffolding whose purpose by and large is to take into account special cases on the boundary of the generalization. Obscuring how and why a theorem works is regarded elegance by pure mathematicians and it is a highly-valued aesthetic of modern mathematics. For the practitioner the scaffolding of generality is but gratuitous complexity which hides useful mathematical machinery and which has to be torn down before the machine can be put to work.

7.3   Purposeful Reading about Astronomy

Astronomy was a popular subject among the readers in experimental philosophy at the Boston Athenæum. One hundred and twenty-four individuals generated 275 charges against forty-four titles. Far and away, the most popular title was *Cosmos: A Sketch of the Physical Description of the Universe* by Alexander von Humboldt with seventy-two charges. Figure 7.5 plots the cumulative charges of all astronomy titles in the upper curve together with the charges of just Humboldt's *Cosmos* in the lower curve. Fairly obviously, the up-tick in astronomy in 1846 is due to the arrival of *Cosmos* on the recent purchases bookshelf.

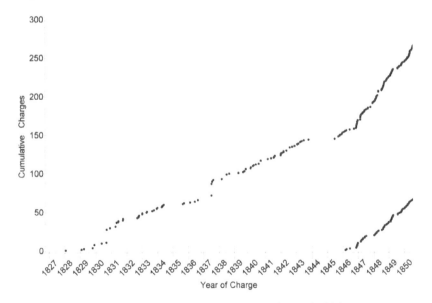

Figure 7.5: Plot of Charges in Astronomy and Humboldt's *Cosmos*

The top reader in astronomy was Solomon Miles with seventeen charges followed by Samuel Goodrich with fifteen charges. Table 7.15 lists the number of readers and charges of popular titles in astronomy and Table 7.16 lists the readers with more than three charges in astronomy.

Author	Short Title	Readers	Charges
Humboldt	*Cosmos*	54	72
Whewell	*Astronomy and General Physics*	18	26
Nichol	*Architecture of the Heavens*	20	24
Laplace	*System of the World*	13	17
Laplace	*Mécanique Céleste*	6	13
Ferguson	*Easy Introduction to Astronomy*	8	11
Bonnycastle	*Introduction to Astronomy*	6	10
Vince	*Complete System of Astronomy*	5	10
Woodhouse	*Elementary Treatise on Astronomy*	5	8
Gregory	*Treatise on Astronomy*	5	6
Biot	*Traité Elementaire d'Astronomie*	3	6
Gummere	*Elementary Treatise on Astronomy*	5	6
Moseley	*Lectures on Astronomy*	6	6
Loomis	*Progress of Astronomy*	5	5

Table 7.15: Popular Titles in Astronomy

Reader	Charges	Reader	Charges
Miles, Solomon Pierson	17	Blagden, George W.	6
Goodrich, Samuel G.	15	Weston, Alden Bradford	6
Cooke, Josiah Parsons, Jr.	9	Quincy, Josiah	5
Jackson, Patrick Tracy	7	Jackson, Lydia	4
Bowditch, Nathaniel	7	Atwood, Charles	4
Nichols, Benjamin Ropes	6	Hammond, Daniel	4
Folsom, Charles	6	Hale, Nathan	4
Tappan, John	6	Parkman, George	4

Table 7.16: Top Readers in Astronomy

7.3.1 Solomon Miles' Astronomy Lectures

The slight up-tick in astronomy in the middle of 1837 is due to Solomon Miles borrowing the astronomy titles listed in Table 7.17.[36]

Author	Short Title	Charges
Woodhouse	*Elementary Treatise on Astronomy*	3
Bonnycastle	*Introduction to Astronomy*	3
Gregory	*Treatise on Astronomy*	2
Biot	*Traite Elementaire d'Astronomie*	2
Norie	*Epitome of Practical Navigation*	1
Biot	*Traite Elementaire d'Astronomie*	1
Ewing	*Practical Astronomy*	1
Gummere	*Elementary Treatise on Astronomy*	1
Laplace	*Mécanique Céleste*	1
Vince	*Complete System of Astronomy*	1
Mackay	*Theory and Practice of Finding Longitude*	1

Table 7.17: Solomon Miles' Charges of Astronomy Titles

Thornton reports that "Bowditch's couriers [of his translation of *Mécanique Céleste* to Europe] included Solomon Miles." [230, p. 221]. Miles could have been brushing up on his astronomy before meeting with the European astronomers but since we don't know when Miles carried Bowditch's work to Europe we can't be certain.

But there is another more plausible explanation for Miles' sudden interest in astronomy. Miles became the Master of Boston English in May of 1823 when George B. Emerson retired. According to Arthur Brayley as recorded in his history of Boston's public schools:

It was in December of that year that the earliest list of the prescribed studies was appointed. This list included Intellectual and Written Arithmetic, by Coburn and Lacroix; Ancient and Modern Geography, by Worcester; History, by Tytler and Grimshaw; Elements of Arts and Sciences, by Blair; Sacred Geography; Reading, Grammar, and Book-keeping; Algebra, by Euler; Rhetoric and Composition, by Blair; Geometry,

---

[36]Miles also checked out Volume 22 of Ree's *Cyclopædia* twice during this period. This volume contains an entry for Mars as well as a lengthy entry on the history of mathematics. Charges of Ree's *Cyclopædia* are however not included in the study.

by Legendre; Natural Philosophy; English Literature and Forensics; Natural Theology, Moral Philosophy, and Evidences of Christianity, by Paley; Practical Mathematics, comprehending Navigation, Surveying, Mensuration, and *Astronomical Calculations with the use of mathematical instruments.*[37] [44, p. 96]

Astronomical calculations were not part of Miles' study at Harvard—recall that his dissertation was a "View of the Seat of Ebenezer Crafts"—so he undoubtedly had to do some homework if he was going to add this branch of mathematics to the Boston English syllabus.

His reading in astronomy in the 1820s must have triggered an abiding interest in the subject as about a decade later Miles volunteered to give a series of six lectures in astronomy in the fourth course of lectures of the Boston Society for the Diffusion of Useful Knowledge. Figure 7.6 is the advertisement for the 1832 lecture series that includes Miles' lectures on astronomy.

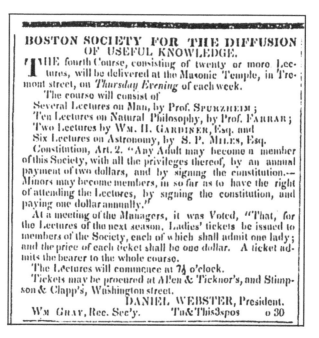

Figure 7.6: Advertisement for Solomon Miles' Astronomy Lectures

---

[37] Current author's emphasis.

The reader may have noticed in passing that the advertisement in Figure 7.6 includes several lectures by Prof. Spurzheim. This is the German physician Johann Gaspar Spurzheim who is known primarily for his work on phrenology. Forty-six of our readers generated sixty-eight charges[38] against five books on phrenology, the most popular of which was Spurzheim's *Outlines of Phrenology*. The advertisement is from the Thursday, November 1, 1832, edition of the *Boston Daily Advertiser & Patriot*. Spurzheim died in Boston nine days later of typhoid. His public autopsy may have had to substitute for his lecture series. He is buried in Mount Auburn Cemetery.[39]

7.3.2   Nathaniel Bowditch's Periodical Reading

One of the first charges[40] recorded in the Athenæum's books borrowed register was that of Nathaniel Bowditch on February 27, 1827. Since Bowditch was the chair of the committee that initiated book borrowing at the Athenæum there may have been a little ceremony when this charge was recorded. Figure 7.7 shows that after dutifully paying his book borrowing fee of $5 and plunking down another $1 for a copy of the catalog, he went home with Volumes 1 and 2 of what appears to be "Mullner." The volumes were due back on March 14.

Figure 7.7: First Charge of Nathaniel Bowditch

On March 14, after returning Volumes 1 and 2 of Mullner, Bowditch checked out Volumes 3 and 4.

There is no "Mullner" in the 1810 catalog. The catalog does list the third edition of *Elements of Mathematics* in two volumes by John Muller. Bowditch had taught himself German in 1818 and in fact had written a joint review of two German works in astronomy, one by William Olbers and the other by Johann Carl Friedrich Gauss, for

---

[38] These charges are not included in the study.

[39] Mr. E.P. Mallory entered a view of Spurzheim's tomb in the Fine Arts department of the 1836 exhibition of the Massachusetts Charitable Mechanic Association. The judges noted "a clever drawing, worthy of notice" but awarded no premium.

[40] ...one would like to imagine the very first. All of Nathaniel Bowditch's charges—in mathematics and otherwise—are listed in Appendix F.

the 1820 edition of the *North American Review*. [230, p. 133–134] Bowditch starts his review of the two books as follows:

> Our object in bringing these works into view at the present time, is not so much to enter into a discussion of the subjects treated in them as to call the attention of the astronomers and mathematicians of our country to some of the improvements in the science, which have been for some time in common use in Germany, but which are hardly known here. [31, p. 260]

At the end of the review there is an editor's addenda gently chiding Bowditch regarding his plans for the publication of his translation of *Mécanique Céleste*:

> We may here be permitted to add what the delicacy of our learned correspondent led him to omit, that a translation of the entire work of the Mécanique Céleste, of de la Place, with a copious commentary, has been completed by the Hon. N. Bowditch; who has not, however, yet been prevailed upon to do honour to himself and to his country, by the publication of so great and arduous a work. [31, p. 272]

Bowditch charged all four volumes of the Athenæum's copy of Laplace's *Mécanique Céleste*. Table 7.18 records the dates of these charges. Each volume was charged twice and these paired charges appear in the same order and at roughly the same time that each volume of Bowditch's translation was going through the press. Bowditch undoubtedly had his own copy of *Mécanique Céleste* so the only thought that comes to my mind is that Bowditch borrowed the pristine volumes to loan them to somebody who was proofreading the galleys of *Mécanique Céleste*. But if this is the case, the unnamed proofreader was reading a text in French to correct a text in English.

Volume	First Charge	Second Charge	Copyright
I	Mar. 11, 1829	June 17, 1829	1829
II	Feb. 13, 1830	June 12, 1830	1832
III	Aug. 6, 1832	Jan. 7, 1833	1834
IV	July 15, 1834	Jan. 16, 1835	1839

Table 7.18: Bowditch's Charges of *Mécanique Céleste*

Of additional interest regarding Bowditch's charges are his readings in scientific journals, the charges for which are listed in Table 7.19.

	Charge Date	Title	Volume(s)
A	Oct. 4 1827	*Philosophical Transactions*	1820
B	Mar. 6, 1830	*Philosophical Magazine*	v.25, v.26, v.27
C	Mar. 9, 1830	*Cambridge Transactions*	v.2
D	Nov. 2, 1830	*Philosophical Magazine*	v.63, v.67, v.68
E	Nov. 2, 1830	*Philosophical Magazine*	N.S. v.1, v.2, v.3
F	Nov. 2, 1830	*Philosophical Magazine*	June 1830
G	Jan. 26, 1831	*Philosophical Magazine*	1827
H	Nov. 18, 1831	*Corr. École Polytechnique*	v.2
I	Nov. 21, 1831	*Corr. École Polytechnique*	v.3
J	June 9, 1832	*Trans. Roy. Soc. Turin*	v.31, v.32
K	Apr. 24, 1833	*Trans. Berlin Academy*	1824
L	Apr. 27, 1833	*Mémoires de Turin*	v.35
M	June 29, 1835	*Trans. Roy. Soc. Edinburgh*	v.11
N	Oct. 7, 1836	*Cambridge Phil. Trans.*	v.5

Table 7.19: Bowditch's Periodical Charges

Very much like the charges of *Mécanique Céleste*, these journal charges span the time that Bowditch was shepherding his translation through the press.

Bowditch made three substantial contributions to Laplace's work above and beyond translating it into English. First, he decompressed and expanded Laplace's mathematics to render it more readily understood. Second, he brought the text up to date so that it included the latest astronomical formulae and observations. And third, as Laplace was acknowledged to be economical in crediting the work of others, Bowditch added citations to uncredited work. The borrowing of sequential blocks of journal volumes could reasonably be taken to be scanning the literature for the two latter reasons. On the other hand and in this context, when only one volume is charged there may well have been particular article of interest to Bowditch himself.

For example, in the first entry in Table 7.19, the 1820 volume of the *Transactions of the Royal Society*, we find "On the Errors in Longitude as Determined by Chronometers at Sea, Arising from the Action of the Iron in the Ships upon the Chronometers" by George Fisher. The amount of iron going into the construction of ships was growing quickly and Bowditch had made a number of contributions to the determination of longitude so the impact of iron on mathematical instruments would undoubtedly have attracted his attention.

Here are my candidates for articles that may have been of interest to Bowditch in some of the charges in Table 7.19:

A George Fisher, "On the Errors in Longitude as Determined by Chronometers at Sea, Arising from the Action of the Iron in the Ships upon the Chronometers"

B, v.25 Ez. Walker, "A simple Way for determining the exact Time of Noon; also a Way to obtain a Meridian Line on a small scale

B, v.26 M. Le Roy, "A Memoir on the best Method of measuring Time at Sea"

C G.B. Airy, "On Laplace's Investigation of the Attraction of Spheroids differing little from a Sphere"

D, v.63 Francis Baily, "On Mr. Babbage's new Machine for calculating and printing Mathematical and Astronomical Tables"

D, v.63 Lieut. Zahrtmann, "On the Mathematical and Astronomical Instrument Makers at Paris"

D, v.67 James Burns, "On finding the Latitude, &c. from three Altitudes of the Sun and the elapsed times"

D, v.68 E. Riddle, "Lynn's new Methods of finding the Longitude"

E, v.1 & G Francis Baily, "Some new auxiliary Tables for determining the apparent Places of Greenwich Stars

E, v.2; & F Encke, "On the Conversion of Right Ascension and Declination Into Longitude and Latitude, and *vise versá*"

E, v.3 "Account of a Paper by Prof. Gauss, intitled [*sic*] '*Disquisitiones generales circa Superficies Curvas*:' communicated to the Royal Society of Göttigen on the 8th of October 1827"

H M. Hachette, "De la projection stéréographique.—Question relative à la sphère céleste.—Sur la transformation des coordonnées"

I M. Puissant, "Sur la détermination de la distance apparente des astres sujets à la parallaxe"

J, v.31 Chevalier Plana, "Note sur un mémoire de M. De-La-Place, ayant pour titre: Sur les deux grandes inégalités de *Jupiter et Saturne*, etc."

J, v.32 J. Plana, "Réfractions astronomiques"

L M. Plana, "Note sur le calcul de la partie du coëfficient de la grande iné galité de *Jupiter* et *Saturne*, qui dépend du carré de la force perturbatrice"

M   John Robison, "Notice regarding a Time-Keeper in the Hall of
         the Royal Society of Edinburgh"

N   G.B. Airy, "On the Latitude of Cambridge Observatory"

7.4   Purposeful Reading about Architecture

There were three architects of prominence among the readers in experimental philosophy at the Boston Athenæum: Solomon Willard, George M. Dexter, and Alexander Parris. The name of another architect, Thomas Dawes, does not appear in the books borrowed registers but a number of books from Dawes' personal library ended up in the Athenæum's collection so Dawes may very well have been a member who just didn't borrow any books.

Architecture is not a category of books borrowed that has been included in this study. Parris borrowed Pocock's *Designs for Churches and Chapels* six times and Pugin's *Examples of Gothic Architecture* four times but neither of these titles are regarded as reading in experimental philosophy. On the other hand, both Parris and Willard borrowed Buchanan's *Practical Essays on Mill Work and other Machinery* which is a title included in the study. The result is that we are seeing architects more as engineers than as artists.

Alexander Parris also donated books from his personal library to the Boston Athenæum, thirty-eight volumes to be exact.[41] Among Parris' seventy-three charges in experimental philosophy there are books that he owned before he borrowed them including Barlow's *Essay on the Strength and Stress of Timber* and Nicholson's *Operative Mechanic*.

7.4.1   Alexander Parris and *Ponts et Chaussées*

Until the mid-1830s Parris' primary business was as an architect-engineer for residential and commercial buildings. Working side-by-side with Charles Bullfinch and Loammi Baldwin, Parris contributed to the design and construction of Boston's Quincy Market and the Cathedral Church of St. Paul and Plymouth's Pilgrim Hall among many other New England architectural landmarks. When residential and commercial commissions dried up in the financial downturn Parris turned to government work, notably to new buildings in Boston's Charlestown Navy Yard, to powder magazines at the Watertown Arsenal, and, most challenging of all, to lighthouses along New England's coastline.

---

[41] The catalog entries for these books all indicate they were donated by Parris but according to Christopher Monkhouse [177, p. 55] some of them were sold to the Athenæum by Parris.

Parris was self-educated and an avid reader so we can imagine him turning pages late into the evening as he got up to speed on these new-departure constructions that weren't at all like graceful townhouses for successful Boston merchants. One of the books he borrowed as he reoriented his career was also one that he had previously donated: Smeaton's folio on the Edystone Lighthouse. [218] Figure 7.8 shows this charge on April 6, 1839.

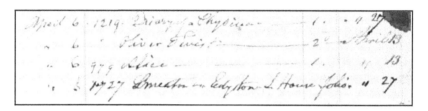

Figure 7.8: Parris' Charge of Smeaton's Folio on the Edystone Lighthouse

Of course, as with all of his donations, we can't be perfectly certain that the Smeaton folio that Parris charged was one-and-the-same folio he donated but in this particular case the likelihood is high.

Parris' most frequent charge was of Robertson's *Mechanics Magazine* which he borrowed twenty-two times. His second-most frequent charge was of a two-volume work by French mathematician and engineer at the famous École Royale des Ponts et Chaussée, Barnabé Brisson. The full title of the first volume is of Brisson's work is:

> *Recueil de 245 Dessins ou Feuilles de Textes Relatifs à l'Art de l'Ingénieur, Extraits de la première Collection terminée en 1820, et lithographiés à l'Ecole Royale des Ponts et chaussées. 1826.*

The full title of the second volume is:

> *Recueil de 239 Dessins ou Feuilles de Textes Relatifs à l'Art de l'Ingénieur, Extraits de la Second Collection terminée en 1825 et lithographiés à l'Ecole Royale des Ponts et Chaussées. 1827.*

Ponts et Chausseées, as it is known in the Athenæum's catalog and in the books borrowed registers, is a folio, measuring roughly 30cm×45cm, consisting primarily of lithographs of French internal transportation projects spanning the period 1797–1825, the period that coincides with the rise and fall of Napoleon and which includes his great port, road and navigation projects both in France and in conquered countries, notably Antwerp in the Southern Netherlands.

Parris borrowed *Ponts et Chauseées* nine times. The dates of these charges as well and the projects he was working on at the time of these charges are listed in Table 7.20.

Charge Date	Volume	Projects
Oct. 26, 1835	v.1	Charlestown Navy Yard Ropewalk and Chelsea Powder Magazine
Dec. 1, 1835	v.2	
May 25, 1836	v.1	
May 29, 1837	v.2	
Dec. 23, 1837	v.2	Whaleback and Saddleback Ledge Lighthouses
Sept. 14, 1839	v.1	
Feb. 5, 1842	v.1	
Apr. 2, 1842	v.2	
Sept. 16, 1842	v.2	

Table 7.20: Parris' Charges of *Ponts et Chausseées*

It is easy to imagine why a book of beautifully detailed engineering and architectural drawings would appeal to Parris even if they didn't have a direct relevance to his projects but it seems to me they did. Volume I, for example, includes the following studies that bear a strong relationship to his government work:

• dry docks at Port de Lorient and Port de Rocheford
• lighthouse at Port de Dieppe
• long, thin marketplace at Ste. Germain (Charlestown Navy Yard ropewalk)
• storeroom for combustible material at Port de Rocheford (Watertown Arsenal powder magazines)

Volume I also contains studies on forges, furnaces, fog horns, water wheels, and pumps. In the same vein, Volume II contains the following:

• navigation beacons at l'Ile d'Aix
• lighthouse on the rocky plateau du Four, west of the tip of the Croisic
• long, thin hangar at Mourillon (Charlestown Navy Yard ropewalk)
• small steam engine invented by MM. Albert and Martin (Lester's steam engine)
• rope spinning machine invented by M. Hubert (Treadwell's speeder)

Parris published what is in a sense an American version of *Ponts et Chausseées*, a 70-page atlas folio entitled *Plans of Buildings and Machinery Erected in the Navy-Yard, Boston from 1830–1840* [188]

7.4.2 Lighthouses, Ropewalks, and Iron Tails

By the end of its first decade of operation the lighthouse marking the entrance to the harbor of Portsmouth, New Hampshire, was falling apart. Built in 1830, the Whaleback Lighthouse leaked in storms and stones were falling out of its walls. In 1839, Portsmouth's lighthouse superintendent, Daniel P. Drown,[42] hired Parris to see what might be done. Parris concluded that the lighthouse had been "constructed without science or workmanship" and submitted a design for a replacement lighthouse but his plan was rejected. Save for patches nothing was done to improve the situation in Portsmouth until 1870 when a new lighthouse was built adjacent to the old one.

Parris' second lighthouse design, the one he created in 1839 for the Saddleback Ledge Lighthouse, was the first of his lighthouse designs to be built and the first lighthouse in the country to be built of finished granite stone. Figure 7.9 shows Saddleback Ledge and a French lighthouse in *Ponts et Chausseées*, that may have been its model.

Phare du Four Lighthouse   Saddleback Ledge Lighthouse

Figure 7.9: Phare du Four and Saddleback Ledge Lighthouses

On a wave-swept granite outcrop in the southeastern entrance to Penobscot Bay, it is a lonely assignment for a light keeper. Saddleback's first keeper, Watson Y. Hopkins, who had moved in with his pregnant wife and seven children in 1839, is quoted in I.W.P. Lewis's report on New England lighthouses as follows:

> I was appointed keeper of this light, December 1839, upon a salary of $450. I live with my family in the tower, which is the only building on the ledge. The tower is in good repair, excepting a leak in the deck on the east side, and the want of any ventilator to the kitchen smoke pipe. I am obliged to

---

[42]...honest.

bring my water from shore, a distance of seven miles....There is a living room and two chambers in the tower, besides a cellar. The copper spout carried round the tower to catch rain water has been so injured by the surf, that it is no longer of any use. The iron railing, which was secured to the rock around the tower, has been all swept away; also, the privy, which was carried away the first storm after its erection. The windows all leak in storms, the shutters having no rebates in the stone work. [164, p. xxx]

It is not hard to understand why most lighthouse keepers were bachelors.

Getting back to Parris' work at the Charlestown Navy Yard, in their 1978 National Register of Historic Places nomination, Edwin C. Bearss and Peter J. Snell list the Navy Yard buildings in Table 7.21 as being designed by him.

Date	Building	Use
1823	21	Commandant's carriage house and stable
1824	n/a	Navy Yard wall
1832	22	U.S.S. Constitution Museum (org. pump house)
1837	34	chemical materials and photo lab
1832[43]	75	timber storehouse or pipe storage shed
1834	58	ropewalk
1836	60	tar house
1837	62	hemp house and rope test lab
1854	38	enlisted men's club[44]

Table 7.21: Buildings Designed by Parris at the Charlestown Navy Yard

The ropewalk complex—buildings 58, 60, and 62, later supplemented by boiler house designed by Joseph Billings and a forklift repair shop—produced rope for the U.S. Navy for 132 years, from the end of the age of sail to beginning of the age of steam. It was also the subject of one of Longfellow's lesser known poems, "The Ropewalk," which begins:

In that building, long and low,
With its windows all a-row,
    Like the port-holes of a hulk,
Human spiders spin and spin,
Backward down their threads so thin
    Dropping, each a hempen bulk.

---

[43] Berass says "or 1848 (available data is conflicting)."
[44] Joint with Joseph Billings.

The engineering community in antebellum Boston was small enough that most practitioners worked together in one context or another at one time or another. For example, just among our readers Nathaniel Bowditch, Solomon Willard, George B. Emerson, H.A.S. Dearborn, William Lyman, Charles C. Starbuck, Alexander Parris, and Daniel Treadwell were all founding members of the Boston Mechanics' Institution. But even in this congenial milieu, the working relationship between Daniel Treadwell and Alexander Parris at the Charlestown Navy Yard strikes me as being closer than the norm. In fact, Leslie Larson in her urban studies report on the Charlestown Navy Yard ropewalk describes their working relationship at the Navy Yard in these words:

> From the earliest conceptual stage for the Ropewalk to its completion Architect/Engineer Parris and Inventor Treadwell worked together closely on the planning and execution of what must be considered a collaborative effort. [158, p. 8]

Figure 7.10 taken from the second volume of *Ponts et Chausseées* is the diagram of a spinning machine component invented by G. Huber. A translation of the text accompanying the diagram is as follows:

> An endless string passes over pulley **A** and over a large pulley that a child[45] turns with a crank. Pulley **A** is fixed to the chassis **BB'CC'**, which thus turns together with it around a horizontal axis **AK**. The wings **DD'** are fixed to the iron cylinder **aa'**, turning by friction in the cylinder **NN'**. The sprocket **P** is fixed to cylinder **aa'**. When this is in place, the chassis **BB'CC'** turns, and the wings **DD**, due to the effect of air resistance, turn less quickly than the chassis, thereby transmitting a rotational movement via the sprocket **P** and the gears to the bobbin **BB'** around the axis **HH'**. From this result two movements, one of which serves to twist the threads, and the other to wind the twisted threads onto the bobbin. [45, Drawing 81]

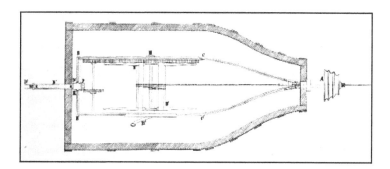

Figure 7.10: Machine à Filer de M. Huber

---

[45] French 'enfant'.

In the Public Documents of the twenty-eighth Congress, dated December 4, 1843, there is a long report on "the culture and manufacture of hemp…so as to produce a sufficiently strong article for naval purposes." where on page 681 we find the following:

> France is indebted to G. Huber for the invention of a small spinning machine, very ingeniously arranged. The trial was first made at Toulon, and crowned with success. The threads spun by this machine are 20 percent improved in strength after being worked up into ropes. The machine is so constructed that the threads are greased or tarred in the hemp, which makes the threads extremely smooth and even. The price of said machine is only from 200 to 250 francs. But, at the same time, it is to be observed that Mr. Huber's machine is only a mechanical assistant, but not a mechanical spinning machine, independent of the assistance and action of the spinner. Therefore, at Toulon are these machines intrusted only to the most skilful workmen. It was proved, by a trial, that from 15 to 20 percent was the gain in threads by this great machine, and it assisted the work men to equalize the first threads on the walk. A point of great importance is that Mr. Huber's machine occupies the least space when in operation. [241, p. 681]

Three years prior to the Senate report, a rope spinning machine invented by Daniel Treadwell completely automated the spinning of rope and in a spin-off at the Charlestown Navy Yard ropewalk produced rope that bested rope spun by "the most skilful workmen" handily.[46]

Given their close working relationship, it is not unlikely that turning the pages of *Ponts et Chausseées* Parris noticed Huber's speeder and pointed it out to Treadwell. One of Treadwell's patents, the one that we will discuss in the following chapter, is for an iron tail which when combined with the top performs the same function as Huber's device.

---

[46] Appendix E lists the five patents that Treadwell was granted for spinning rope from hemp.

We see that the artificers, that never dreamed of the Epicurian philosophy, have accommodated mankind with a multitude of useful inventions.

A Proëmical Essay
Robert Boyle

Chapter 8

Patents and Inventors

8.1   Robert H. Eddy and Sulphuric Ether

On June 18, 1847, Robert H. Eddy borrowed *The Life of Edward Jenner* by John Baron. It was due back on June 22. Eddy was a civil engineer by training but executed a mid-life career change to patent solicitor in the mid-1830s. His reading, including even his recreational reading, didn't stray too far from technical matters so this charge attracts one's attention.

Eight months before he checked out *Jenner's Life* Eddy had successfully prosecuted the granting of patent US 4,848, issued on November 12, 1846, for an "Improvement in Surgical Operations" on behalf of its inventors, William T.G. Morton and Charles T. Jackson. The first paragraph of US 4,848 reads in part:

> [W]hereby we are enabled to accomplish many, if not all, operations, such as are usually attended with more or less pain and suffering, without any or with very little pain to or muscular action of persons who undergo the same.

This medical miracle was accomplished by using sulfuric ether to sedate the patient during the operation. The patent started to get push-back from the medical community immediately after its issuance due in no small part to Morton's aggressive efforts seeking royalties for its practice.

It is pure conjecture, of course, but when Eddy borrowed *Jenner's Life* he may have been wondering how Jenner had been able to profit from his discovery of a small-pox vaccination without having obtained a patent. Another possibility is that Eddy had spotted a reference to Baron's book in Edward Warren's polemical tract, *Some Account of the Letheon: or, Who is the discoverer?* [246, p. 74]. Morton's co-inventor, Charles T.

Jackson, had begun to claim he had given Morton a hint about anesthetic properties of sulphuric ether and thus he, Jackson, should be credited as being the sole discoverer. Morton's counter-argument was that it is the actualization of an inventive idea, not hints about its possibility, to which credit accrues. In Jenner's case, a "young country-woman" had said to him "I cannot take the disease, for I have had cow-pox."[10, p. 122] and yet it was Jenner not the young country-woman who was and still is credited as being the sole discoverer of the small-pox vaccination.

The 1836 Patent Act[1] called for the examination of patent applications by federally-employed patent examiners prior to and as a condition of a patent being granted. The examiners were charged to ensure that the invention described in a patent application was "new, useful, and non-obvious" in the words of the Act. When the first patent act was passed in 1790, patent examination was the duty of the Secretary of State, the Secretary of War, and the Attorney General. To nobody's surprise in retrospect, Thomas Jefferson, Henry Knox, and Edmund Randolph had more pressing demands on their time than to determine if Samuel Mulliken's machine for raising nap on cloths was worthy of patent protection. The hastily passed fix, the 1793 Patent Act, simply said that patents were to be granted upon proper payment of the necessary fees and punted the determination of the validity of patents to courts, lawyers, judges, and juries. This got Jefferson, Knox, and Randolph off the hook but but it didn't address the question of who was to determine if Mr. Mulliken's wonderful machine should be granted a patent in the first place.

It wasn't until the passage of the 1836 Patent Act that this second problem was addressed. This legislation stipulated that all patent applications were required to include full descriptions of an explicitly prescribed format and that...

> ...before any inventor shall receive a patent for any such new invention or discovery, he shall deliver a written description of his invention or discovery, and of the manner and process of making, constructing, using, and compounding the same, in such full, clear, and exact terms, avoiding unnecessary prolixity, as to enable any person skilled in the art or science to which it appertains, or with which it is most nearly connected, to make, construct, compound, and use the same. [148, p. 159]

Note carefully the phrase "avoiding unnecessary prolixity." Boyle used almost exactly these words in describing the experimental essay.

The 1836 Patent Act also mandated that patents be examined. While the Act was silent on the qualifications of the examiners, it made clear that henceforth getting a patent was going to be more than just paying filing fees:

---

[1]Ch. 357, 5 Stat. 117 (July 4, 1836) An Act to promote the progress of useful arts, and to repeal all acts and parts of acts heretofore made for that purpose.

> But whenever, on such examination, it shall appear to the Commissioner that the applicant was not the original and first inventor or discoverer thereof, or that any part of that which is claimed as new had before been invented or discovered, or patented, or described in any printed publication in this or any foreign country, as aforesaid, or that the description is defective and insufficient, he shall notify the applicant thereof, giving him, briefly, such information and references as may be useful in judging of the propriety of renewing his application, or of altering his specification to embrace only that part of the invention or discovery which is new. [Section 7, Patent Act of 1836]

One imagines that patent solicitors were available for hire in Massachusetts from the time that Samuel Winslow was given exclusive right to practice his new method for making sails by the Massachusetts General Court in 1641 but until the 1836 Patent Act was passed patent solicitation was more a matter of lobbying than it was a matter of making a cogent technical argument.[2] With the passage of the 1836 Patent Act somebody who presumptively understood the technology of a patent application could reject the application on technical as well as procedural grounds. This greatly altered the nature of patent solicitation, very much in Robert H. Eddy's favor.

Boston business directories didn't start listing patent agencies or patent solicitors as a separate category for over a decade after the 1836 Act was signed into law but when they finally did the length of the lists, as tabulated in Table 8.1, were small. Whether this was because there were other law specialities that were more rewarding or because the amount of technical expertise required to practice patent law was daunting, it certainly wasn't for lack of business.

Year	Number	Year	Number
1848	3	1855	9
1849	3	1856	6
1850	3	1857	11
1851	3	1858	8
1852	2	1859	7
1853	3	1860	4
1854	5		

Table 8.1: Boston Patent Solicitors and Agencies, 1848–1860

---

[2]Winslow's patent is generally regarded to be the first granted in the New World. The first patent ever granted was to Franciscus Petri in 1416 for a fulling machine. [254, p. 106]

We met Robert H. Eddy with his civil engineering hat on in Chapter 3 where he was one of the long line of experts hired to advise one of the many committees trying to bring water to Boston. Eddy's name is, in fact, attached to a number of large engineering projects in and around Boston. For example, he surveyed and laid out the Park Hill subdivision in Worcester. But at the same time he was riding herd on his engineering projects Eddy was ramping up his new career as a patent solicitor. Writing in the royal third person in his family genealogy, Eddy recalls:

> [H]e reported on Spot and Mystic Ponds as good sources of [water] supply. Subsequently he abandoned engineering and became a Solicitor of Patents, being the first to establish the business in this country. Up to 1880 he has been a very extensive practitioner, and, in all probability, has procured for inventors more patents than any other person. [78, p. 206]

According to a blurb in an 1885 Boston business directory:

> Mr. R. H. Eddy, the well-known solicitor of patents, established himself in business in the year 1832, and is believed to have been the first regular solicitor to appear before the United States Patent Office in behalf of an inventor…He occupies an entire floor in a handsome building.

While his office at 76 State Street may have been handsome, Eddy was by no means the first solicitor to appear before the patent office. William Blagrove beat Eddy out by over a decade when he, Blagrove, opened an office in Washington, D.C., in 1819 to prosecute patents for his clients. [73, p. 98]

The first appearance of Eddy in the patent record is as a witness two patents issued on January 13, 1835: 8,586X[3] "Machine for Breaking the Husks of Coffee Berries" issued to Isaac Adams and 8,591X "Machine for Breaking the Outer Husks of the Coffee Berry" issued to Thomas Ditson.[4] Witnesses on early patents testify to the veracity signature of the inventor rather like a notary and not to the fact that the inventor had invented anything new, useful, and non-obvious. Thus, that Eddy was a witness on a patent does not necessarily mean he was the patent's solicitor but that he was a budding patent solicitor at the same time bumps up the likelihood. Figures 8.1 and 8.2 are the drawings from these two early Eddy patents.[5]

---

[3]See the footnote on page 59 for a description of X patents.

[4]Ditson may have been the Billerica farmer whose tar and feathering by a band of Tories popularized the song "Yankee Doodle."

[5]The *Boston Mechanic* was not impressed with either of these patents: "Both these patents appear to have little claim to originality, and about as little to superiority over those invented for hulling other materials."[84, p. 169]

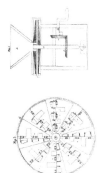

Figure 8.1: Drawing for
Isaac Adams'
Patent 8,586X

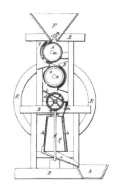

Figure 8.2: Drawing for
Thomas Ditson's
Patent 8,591X

Eddy was himself granted two patents: US 3,582 "Lamp-Cap" issued on May 10, 1844, and US 115,722 "Improvement in Suspenders" issued on June 6, 1871. The lamp-cap patent is an improvement on an invention of Joseph Benson for whom Eddy had previously obtained patent US 3,454 "Oil Can" on February 28, 1844. With regards to the possibility of a raised conflict-of-interest eyebrow, Eddy signed over his rights for his "improvement in oil-feeders...for which Letters Patent were granted to Joseph Benson" patent to Deming Jarves and the New England Glass Company. At the beginning of US 3,582 Eddy comes clean:

> Not being desirous owing to circumstances connected with my profession of availing myself of any advantages thereof beyond what I may receive as compensation for preparing the necessary papers, etcetra, for securing the same by Letters Patent and the legal representatives of the said Benson....I assign all my right, title, and interest in the said invention to Deming Jarves and the New England Glass Company.

As we shall see shortly, assigning rights does not preclude the existence of a side agreement for sharing revenues.

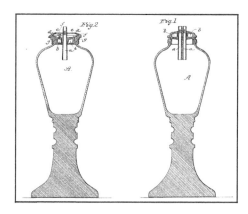

Figure 8.3: Drawings for Robert H. Eddy's Patent US 3,582

The question of whether or not Eddy "procured for inventors more patents than any other person" as he claimed in his memoirs is untested. New York's Munn & Co. certainly dominated patent prosecution in the second half of the nineteenth century. [73, pp. 173–76] But that Eddy was busy filing patent applications in and around New England in the middle of the nineteenth century is beyond doubt. There are over 400 patents with Eddy's signature in Google Patents.

And it is a lingering memorial to Eddy's diligence that some of the patents he filed are still being cited as prior art in patent applications today. Frank Munroe's whip-sockets patent, US 143,417, prosecuted by Eddy and granted on Oct. 7, 1873, was cited by US 791,0067 "Sample Tube Holder" granted on March 22, 2011. And Noah Lindley's August 25, 1842, patent on the construction of churns that Eddy represented was cited by US 830,3166 "Food Flipping and Turning Spatula" granted on November 6, 2012.

The story of the discovery of the surgical use of ether and the priority fights that followed the granting of US 4,848 is well-told by Julie Fenster in *Ether Day: The Strange Tale of America's Greatest Medical Discovery and the Haunted Men Who Made It* [102] as well as in other books and hundreds of journal articles. There is no need to revisit this tortured and twisty tale here. There is, however, an aspect of this history that loops back to Eddy's interest in *The Life of Jenner*.

Fenster notes that "On October 1, the day after his first successful etherization, [Morton] was already conferring with R.H. Eddy, Esq., regarding the possibility of a patent on ether." Shortly thereafter Charles T. Jackson visited Eddy to press his claims to have tipped off Morton to the anesthetic properties of sulphuric ether. Jackson didn't want his name associated publicly with use of ether should it fail but at the same time he did want a piece of the action should it succeed. Eddy must have been expecting Jackson's visit. Reminding Jackson of how he, Jackson, had been cut out of credit as well as monetary reward for the invention of the telegraph by not attaching his name to Samuel Morse's US 1,647, , Eddy proposed adding Jackson's name to the ether patent application. According to Fenster:

> Dr. Jackson agreed to apply for the patent with Morton, and to split the profits: Morton was to receive 65 percent of the revenues; Eddy, 25; and Jackson, 10. [102, p. 99–100]

Thus, just three weeks after Ether Day and merely four days before the filing of the patent application Eddy was, at least in Fenster's telling, in possession of rights to twenty-five percent of the profits from licensing the patent—more than one of the putative inventors—should it be granted. Moreover, he was able to negotiate further business arrangements with respect to this royalty stream without consulting Morton. [102, p. 100]

Eddy's side agreement with Morton regarding a share in the proceeds from the patent did not go unnoticed by Jackson's supporters during the fray over credit for the invention:

> Now Mr. Eddy will not deny that he had an interest in Mr. Morton's patent; he will not deny that he expected to realize a large profit from the discovery in Europe; indeed, it is admitted, in the argument of Mr. Dana, that he was interested to the amount of fifty percent in the European rights,[6]

This suggests that Eddy and Morton were working together long before Ether Day. Whether or not this is so, it is more than a little curious that Eddy was able to hive off twenty-five percent of what both of them fantasized to be a multi-million dollar revenue stream just for filing a two-page patent application.

Fenster goes on to take Eddy to task for even filing the patent application in the first place:

> [Morton] had received flimsy legal advice. A known substance, such as sulfuric ether, could not meet the legal definition of an invention of design. A more discerning lawyer might have advised Morton that patent protection, even if granted, could not hold. A more dispassionate lawyer might have told him that retaining commercial rights over a common substance—cornering the market on it—was out of the question. [102, p. 158]

There is no written record of what advice Eddy gave Morton. He may well have advised Morton on the likelihood of receiving a patent and on the chances of realizing

---

[6]See [86, p. 505] and [166, p. 15].

any profits from the patent should it be granted, although having cut himself in on the latter may have contoured his words on the former. A patent attorney is not the patent's examiner nor is he an officer of the Patent Office in the way that a member of the bar is an officer of the court. If his client insists on filing a patent application,[7] it is up to the examiner not the solicitor to determine if the patent should or should not be granted.

Further to Fenster's objection, Eddy was not seeking to patent sulfuric ether. He sought to patent one of its uses. There is only one claim in US 4,848:

> What we claim as our invention is—
>
> The hereinbefore-described means by which we are enabled to effect the above highly-important improvement in surgical operations—viz., by combining therewith the application of ether or the vapor thereof— substantially as above specified.

We know the patent was issued so the patent office presumably understood that it was for a use of sulfuric ether not for sulfuric ether itself. We also know that the patent was ultimately declared invalid but not for the reason that Fenster cites. In invalidating US 4,848 the court wrote:

> Neither the natural functions of an animal upon which or through which it may be designed to operate, nor any of the useful purposes to which it may be applied, can form any essential parts of the combination, however they may illustrate and establish its usefulness. Motion for a new trial denied.[8]

In any event, it was the medical community as much as the patent office that denied Eddy, Morton, and Jackson their windfall. As a matter of professional ethics, followers of Hippocrates, including dentists, are not supposed to restrict access to or even profit from advances in the medical sciences. Morton suffered many slings and arrows for applying for his patent. He ultimately abandoned any hope of financial gain from it and died impoverished.

But back to Baron's *Life of Jenner*. Eddy may well have seen the push-back from the medical community coming when he borrowed the book. Discussing how the government might reward Jenner financially for his discovery of smallpox vaccine, Baron describes an alternate to granting a patent:

---

[7] ...even for a perpetual motion machine.

[8] Morton v. New York Eye Infirmary, Dec. 1, 1862. The court noted in tossing out the patent that "[D]istinguished surgeons of the city of New York agreed in ranking it among the great discoveries of modern times; and one of them remarked that its value was too great to be estimated in dollars and cents." Ouch.

Mr. Fuller thought the larger sum (20,000£) due to Dr. Jenner—more especially as he could look to *no remuneration* by *patent*.[9] [10, p. 508].

The debate over credit for the discovery of the anesthetic use of sulphuric ether split the medical community into two camps; those supporting Morton's claim and those supporting Jackson's. One of our readers, Augustus A. Gould, was firmly in Morton's camp during the priority scrum. Gould had personally witnessed the dawn of the invention and provided significant mechanical engineering assistance to its practice. In fact, as will be recounted later in this chapter, Gould received a patent of the inhaling apparatus used to administer ether.

Another of our readers, Dr. Josiah Flagg, Jr., was just as firmly in Jackson's camp. I have found nothing of direct relevance in Flagg's Athenæum reading that might bear on the Ether Day dispute but there is one borrowed book listed on his register pages that is of at least passing note. On April 10, 1832, and again on August 21, 1833, Flagg checked out Volume II of Hamilton's *History of Medicine*. Starting on page 274, Hamilton retells the Jenner discovery story but rather than arguing that credit be shared with the nameless young country-woman, Hamilton opines that credit should be shared with one Mr. Cline:

> The merit which thus indisputably belongs to Mr. Cline of having vaccinated the first patient in London, has been attempted to be taken from him by the claim of another practitioner, which is far from resting on an equally solid foundation.…if the tooth of calumny still makes puny efforts to corrode the adamantine pillar of Jenner's well earned fame, her efforts only resemble those of the viper against the file—reacting upon herself, and making her the victim of wounds designed for another. [125, pp. 275–276]

Strong stuff. Now, it is highly unlikely that Flagg brought this passage in Hamilton back to mind a decade on—although, puny efforts of the tooth of calumny is a phrase that might linger on in a dentist's memory—he certainly was familiar with an author who had expressed strong views on the matter of allocating credit for an invention.

8.2   Purposeful Reading by Inventors

William Rosen in his book *The Most Powerful Idea in the World: A Story of Steam, Industry, and Invention* [205] argues convincingly that Section VI of England's 1623 Statute on Monopolies[10] was the most powerful idea in the world because it triggered

---

[9]Emphasis Baron's.

[10]Formally the "Act concerning Monopolies and Dispensations with penall Lawes and the Forfeyture thereof."

the Industrial Revolution. The section on the granting of monopoly rights to a "first and true inventor" was the only monopoly that was not precluded by the statute. The industrial projectors patronizing the Boston Athenæum were keenly aware of the power of patents, both the financial rewards they yielded if you held one and the expense of licenses and infringement proceedings if you didn't. Consequently, up-to-date knowledge of the patent landscape—existing patents, prior art, infringement law suits, etc.—was just as necessary for a venture capitalist as a copy of Bowditch was for a sea captain.

A patent application is legally obliged to contain sufficient detail so that a person versed in the patent's technology can successfully construct a working instance of the patented device in the same way a reader of an experimental essay can replicate the experiment. In the words found in one of the patents we will consider below, a patent shall "enable others skilled in the art to make and use my invention."[37] This is after all the *quid pro quo* of the patent system. The inventor gets a limited term exclusive right to practice the patent in return for enabling anyone to practice the patent using nothing more than the text of the application after the patent expires.

Six titles recorded in the books borrowed registers are specifically about inventions and patents. Two of the four, *History of Inventions* and *Century of Innovations* are history books. The other four are periodicals devoted to reporting on patent news of the day. The author, short title, and number of readers and charges for all six are listed in Table 8.2. Figure 8.4 is the cumulative charge plot of the six titles.

Editor/Author	Short Title	Readers	Charges
Newton	*Repertory of Patent Inventions*	27	60
Beckmann	*History of Inventions*	26	38
Worcester	*Century of Innovations*	4	4
Pritchard	*English Patents Granted*	2	2
Treuttel	*Archives des Inventions*	2	2
Ellsworth	*Digest of United States Patents*	2	2

Table 8.2: Titles of Books and Periodicals on Patents

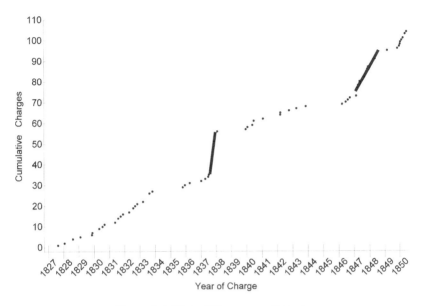

Figure 8.4: Plot of Charges on Patents

Period	Start	End	Readers	Charges	Interest
	Jan. 1827	Dec. 1850	57	106	12
A	Mar. 1837	Apr. 1837	2	20	153
B	Feb. 1847	July 1848	17	22	38

Table 8.3: Periods of Elevated Interest in Patents

The jump in 1837, Period A, marks Robert H. Eddy's career change from civil engineer to patent solicitor. Eddy speed-read his way through nineteen volumes of the *Repertory of Patent Inventions* between March 18 and March 22 of that year.

In addition to the four journals listed in Table 8.2 that are devoted to patents, three additional journals recorded in the books borrowed registers contained extensive reports and commentary on patents granted as a regular feature. The titles of these three journals along with counts of their readers and charges are listed in Table 8.4.

Editor	Short Title	Readers	Charges
Robertson	*London Mechanics Magazine*	37	103
Silliman	*American Journal of Science*	38	98
Jones	*Franklin Journal*	38	92

Table 8.4: Periodicals Reporting on Patents

Figure 8.5 is the cumulative charge plot of the journals in Table 8.4. The solid line tracks charges of *London Mechanics Magazine*, the dashed line tracks *American Journal of Science*[11], and the dotted line shows charges of the *Franklin Journal*.

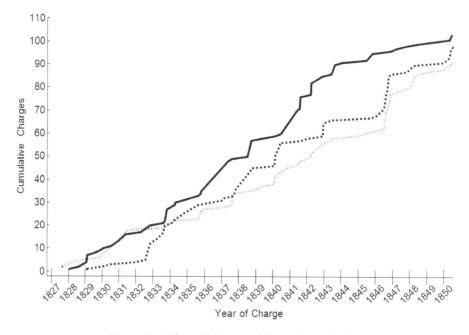

Figure 8.5: Plot of Charges of Patent Periodicals

The bulk of the content in these three journals consisted of short articles about technical topics of the day so a charge of any one of them cannot be counted unequivocally as an interest in patents. But anyone tracking the issuance of patents or contemplating

[11] ...usually referred to as *Silliman's Journal*.

applying for their own would certainly consult these three periodicals regularly for the latest news on patents in particular and technology in general.

Of the three, the *Franklin Journal* carried the best and most exhaustive coverage of American patents. Dr. Thomas P. Jones[12] was its editor from the first issue of the *Franklin Journal* in January 1826 until his death in 1848. He was also fleetingly the Superintendent of the Patent Office before the 1836 Act was passed. When the act went into effect, he was appointed to one of the two patent examiner positions it stipulated.

Jones published verbatim excerpts from patent applications in the *Franklin Journal* and frequently added editorial remarks about the patent's technology and its claims. Understanding how an ex-superintendent of the Patent Office and one of the current patent examiners analyzed patents would obviously be more than just a little useful in crafting a new patent application. This is doubtless one reason why the *Franklin Journal* enjoyed such an avid following among inventors and patent solicitors of the day. To be sure, Jones was well aware that he was wearing two hats. He wrote in an issue of the journal published shortly after the 1836 Act was passed:

> Under this law, there are two examiners appointed, of whom the Editor is one, and it will, therefore, be manifest that the tone of his animadversions, must be modified by the existence of these circumstances. [149, p. 327]

In addition to reporting on patents granted, the three journals in Table 8.4 also carried reports of infringement proceedings.[13] Testimony in these cases often revealed prior art that was unencumbered by patent protection as well as additional patents that might need to be cited, licensed or worked around. Testimony by the inventor and by experts on both sides of the pleadings could also reveal the thinking that went into the invention. All of this would obviously be invaluable to promoters and practitioners whether or not they were contemplating filing a patent application of their own.

---

[12]For a history of the *The Journal of the Franklin Institute* see [192]. For biographical information on Jones related to the journal see [106]. Jones was trained as a physician, held the chair of the chemistry department at Columbian College, and he…

> …wrote the preface to [Oliver Evans'] *The Young Mill-wright and Miller's Guide* after the 4th edition and also carefully edited these works using his personal judgement to correct only those theoretical views advanced by Evans, which by the passage of time and increased knowledge might be disputed. [11, p. 264].

[13]When it comes to patent infringement, inventors sort themselves into two mutually exclusive camps. Some such as W.T.G. Morton and Oliver Evans become obsessed with fighting real and imagined infringements of their patents. Others such as Daniel Treadwell simply walk away to invent another day. Morton was undoubtedly well aware of Charles T. Jackson's habit of challenging patents to which he believed he had contributed: gun cotton and the telegraph to name just two. This may be one reason why Eddy convinced Morton to add Jackson's name to US 4,848 and subsequently convinced Jackson to sign over his rights to the patent to Morton…and silently to Eddy himself as it turned out.

8.3   Patent Holders among the Readers in Experimental Philosophy

Twenty-four of the 702 individuals whose names are recorded in the books borrowed registers were granted collectively ninety U.S. patents during the period of the study. Some of the patents, such as the one granted to Augustus A. Gould for an "Apparatus for Inhaling Ether, &c." and the many granted to Erastus B. Bigelow on the weaving of rugs, come as no surprise. Josiah Flagg's patent on "Locomotive Spark-Arrester and Smoke-Conductor" might be a little less expected.

Matching a name on a register page with a name on a patent application with the intention of concluding that they refer to one and the same individual comes with all of the well-known uncertainties. In the case at hand, however, there are some scraps of ancillary information that can tip the scale one way or the other.

The city in which the individual resided when a patent was filed was required to be included in the patent application. In our case, if the city of residence is Boston or a nearby town then it is feasible that the inventor was also a member of the Boston Athenæum. If the city of residence was Cincinnati, then it is less likely. The Erastus B. Bigelow of West Boylston, Massachusetts, whose name appears on patents having to do with weaving rugs is highly likely to be the E.B. Bigelow who is allocated a page 485 in Volume IV of the books borrowed register. On the other hand, the Samuel G. Dorr who was granted patent 46X on October 20, 1792, for a "machine called the 'wheel of knives' shearing and raising the nap on cloths" and lived in Albany, New York, is unlikely to be the Samuel Dorr who was allocated page 128 of Volume I and borrowed mostly histories and novels.[14]

The degree of overlap between the subject matter of a patent and the books that the individual was taking home can also contribute to resolving to a name match. The Augustus A. Gould of Boston, Massachusetts, who was granted a patent for an apparatus for inhaling ether is almost certainly the Athenæum reader Augustus A. Gould, M.D., who borrowed Muller's *Physiology* along with numerous other medical books.

Keeping all of these cautions in mind, Table 8.5 lists the names of the twenty-four Athenæum readers who were granted one or more US patents. The counts in the table reflect the number of patents the individuals were granted between 1790 and 1850 but do not include reissued patents.

---

[14]...not saying that an inventor of a wheel of knives might not be interested in history.

Reader	Patents	Reader	Patents
Alger, Cyrus	5	Gould, Augustus A.	1
Bigelow, Erastus B.	20	Hale, Nathan	2
Boyden, Uriah A.	3	Higginson, Henry	1
Chandler, Abiel	1	Homer, Charles S.	1
Curtis, Caleb	1	Hurd, Joseph Jr.	11
Dexter, George M.	1	Jackson, Patrick Tracy	1
Jackson, Charles T.	1	Lyman, William	1
Eddy, Robert H.	1	Moody, Paul	10
Flagg, Josiah F.	2	Savage, James S.	1
Forbes, Robert B.	1	Robinson, Enoch	6
Foster, Leonard	6	Treadwell, Daniel	10
Goodrich, Samuel	1	Whipple, John A.	2

Table 8.5: Patent Counts for Athenæum Readers

Table E.1 in Appendix E gives the title, patent number and grant date of each patent for each of these twenty-four readers. Table E.2 in the same appendix lists seventeen additional couplings between a register name and a patent name that were regarded as not being sufficiently strong to be included in Table 8.5.

The first Athenæum reader to be granted a U.S. patent was Dr. Josiah Foster Flagg. Flagg and his son, Josiah, Jr., were granted patent *1,307X* on May 17, 1810, for an "*elastic lamp reflector or elastic tubes for Argand's lamps.*" An Argand lamp, a small oil lamp used primarily for home lighting, was invented and subsequently patented in France by Aimé Argand in 1780. The patent document describing Flagg's improvement on Argand's lamp is not among those which have been recovered after the 1836 patent office fire so we don't know how Flagg's reflector and tubes improved Argand's lamp.

Of course, a patent wasn't the only way to gain recognition for one's inventive prowess. The exhibition of the Massachusetts Charitable Mechanic Association held in 1836, 1839, 1841, 1844, and 1848, for example, was explicitly held to be a showcase for innovation. Table 8.6 list some entries of readers at the Athenæum in this event.

Reader	Year	Entry
Alger, Cyrus	1836	Two iron and one brass pieces of heavy ordnance
Alger, Cyrus	1839	Two Brass Six Pounders
Alger, Cyrus	1839	Medallion of St. John
Alger, Cyrus	1839	Malleable Cast Iron Plough
Alger, Cyrus	1841	Two Light Brass Six Pounder Chambered Navy Guns
Alger, Francis	1839	One Six Pounder, mounted
Davis, Isaac P.	1836	Carved Shell Combs
Eddy, Robert H.[15]	1836	Mechanical Drawings
Eddy, Robert H.	1836	Model of Swing Draw, for rail-road and other bridges
Hoit, A. G.	1841	Two portraits. *A Silver Medal*[16]
Merrill, Amos B.	1839	Specimens of Dried Plants from Hampden
Loring, Josiah	1836	One set eighteen inch Globes, and one set twelve inch Globes
Loring, Josiah	1839	A pair of Eighteen Inch Globes
Parris, Alexander	1841	Model of a Beacon
Parris, Alexander	1841	One Drawing of a Cast Iron Beacon
Parsons, Theophilus[17]	1844	One Specimen of Bird Stuffing
Shattuck, Lemuel	1841	A Family Register
Shattuck, Lemuel	1844	One Specimen of Book Composition
Warren, John W.	1839	Bathing Tent. *A Silver Medal*
Whipple, John A.	1841	Daguerreotype Portraits
Whipple, John A.	1848	Specimens of Daguerreotypes

Table 8.6: Example Reader Entries in the MCMA Exhibitions

---

[15] Eddy's father, Caleb, exhibited tonsil instruments in 1839.

[16] The judges were moved to note "We look in vain, however, among the works of the artists of the present day, for that brilliant and true coloring so conspicuous in the portraits of Stuart."

[17] Parsons along with Thomas Sherwin were judges in the Mathematical, Philosophical, Optical, and School Apparatus department at the 1848 exhibition.

8.3.1   Cyrus Alger and Casting Iron

The second Athenæum reader to receive a patent was Cyrus Alger. Alger was issued patent number 1,483X on March 30, 1811, for new method of "Casting Iron Rolls." Figure 8.6 is the drawing on the first page of Alger's patent.

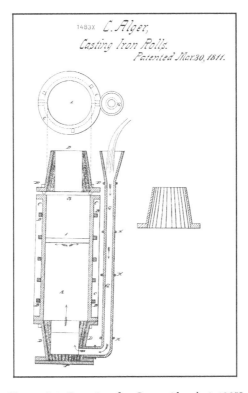

Figure 8.6: Drawing for Cyrus Alger's 1,483X

The challenge in making rollers for rolling hot metal was to make them so smooth that the metal sheets and plates against which the rollers pressed were themselves smooth after being rolled. Prior to Alger's invention, a freshly-cast roller had to be turned on a metal lathe to even its surface before it could put into use. This was a slow and labor-intensive step in the roller manufacturing process. Alger's patent discloses a method for casting rollers that were sufficiently smooth right out of the mold.

Alger's company, the South Boston Iron Company, was renowned for supplying high quality cannons including the massive Columbiad coastal defense mortar used in the War of 1812. Figure 8.7 is an advertisement for Alger's foundry.

Figure 8.7: Advertisement for Cyrus Alger's Foundry

8.3.2   Augustus A. Gould and Inhaling Ether

Augustus A. Gould's lone patent, US 5,365, "Apparatus for Inhaling Ether" granted on November 13, 1847, was joint with W.T.G Morton. Gould assigned his rights to the patent to Morton. Robert H. Eddy was a witness to the Gould/Morton patent as he was the Morton/Jackson patent. It is not recorded if Eddy had a financial interest in US 5,365 as he did in US 4,848, but a convenient device for administering ether was obviously needed for ether to be a success, medically and financially.

Figure 8.8 is the drawing from Gould's patent. The open end of the tube **B** on the left-hand side of the diagram was placed in the patient's mouth. When the patient inhaled, air containing whatever vapor was coming off the sponge **C** would come through one-way value **b** at the right-hand end of the mouthpiece. When the patient exhaled, air would flow out of the one-way valve **e** at the top of the mouthpiece. The stopper whose handle is at the very top of the diagram was used to control the dripping of the fluid in the upper reservoir **D** down onto the sponge **C** in the lower reservoir. Air flowed into the lower vessel through the one-way valve **k** on the right-hand side of the lower reservoir.

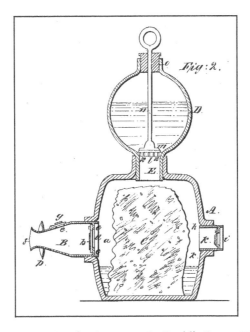

Figure 8.8: Drawing for Augustus A. Gould's Patent US 5,365

There were five claims in the Gould/Mortan patent for their inhaler:

1) the mouthpiece, **B**
2) the upper reservoir, **D**
3) the flange at the patient's end of the mouthpiece, **p**
4) the lower reservoir, **A**
5) the sponge in the lower reservoir, **C**

Morton was a roomer in Gould's home running up to Ether Day. Fenster reports that Gould recalled in his memoir that on the evening before the legendary October 16 operation:

> Dr. Morton called to ascertain about the problematic injurious effects of ether, and what articles might be used. I answered; and in the course of the conversation I asked him how he gave it. He told me that he put a sponge in a globe saturated with ether, and drew the vapors through a tube attached; breathed out and in through a tube attached. I suggested that the application of valves, to prevent breathing back the air into the globe, would be desirable, and sketched a plan. He said "that is it; that is just it. I will have it for tomorrow." [102, p. 9–10]

There is nothing in Gould's reading as recorded in the books borrowed registers that relates directly to his patent. US 5,365 is likely to be more the product of time spent working in the laboratory and the surgical amphitheater than to reading at the Athenæum.

It may be noted in passing that Gould read far more at the Athenæum than is recorded in the books borrowed register. Jeffries Wyman in his biographical memoir of Gould tells of Gould's early work in the library:

> Until his [medical] profession could yield him a support, he must go out of it, and did, to earn the necessaries of life. To this end he undertook burdensome tasks; one of them, the cataloguing and classification of the fifty thousand pamphlets in the library of the Boston Athenæum, was herculean, as any one may see who will take the trouble to look over the four large folio volumes[18] he wrote out, monuments of his patient industry, for which he received fifty dollars. [255, p. 95]

Wyman's memoir is devoted almost entirely to Gould's contributions to conchology. The memoir includes an exhaustive bibliography but strangely fails to mention Gould's patent. Today's patents are written by patent lawyers and thus represent the inventor's ideas encoded in patentese and therefore not reliably in the inventor's own words. Up until the middle of the nineteenth century, however, patent applications were written by the inventor and, in the case of most of the X patents, in the inventor's own hand. Whether the words are crafted by the patentee or not, a patent must by law adhere closely to the patentee's thinking and thus one would think would be included in his bibliography, perhaps as a translation into a foreign language.

8.3.3   Uriah A. Boyden and Swirling Water

Recall from Chapter 3 Euler's Rule for water wheel design:

> For maximum effect the water must act on the wheel without impact and leave it without velocity.

In short, no splashing. That wheelwrights understood Euler's Rule is evidenced in the water wheel patents they applied for and were granted. As is so often the case with a game-changing innovation, the idea of the water-runs-through-it hydraulic turbine was in the air long before Ellwood Morris built a copy of Fourneyron's turbine at the Rockland Cotton Mills in 1843.

There are 288 entries under "Water Wheels" in Edmund Burke's list of United States patents issued between 1790 and 1847. [49, pp. 230–235]. Table 8.7 is a categorization of sixty-five of these that foretell the coming of the hydraulic turbine.

---

[18]See [25].

Category	Patents
water wheel, reacting/reaction	26
water wheel, horizontal	21
water wheel, applying water to	7
water wheel, tub	6
water wheel, flutter	3
water wheel, conical	2

Table 8.7: Counts of Pre-Turbine Water Wheel Patents, 1790–1847

The earliest of the proto-turbine patents on Burke's list is *21X* granted to James Macomb of Princeton, New Jersey, on June 26, 1791: "Horizontal Hollow Water Wheel for Mills." Another patent that could be taken as a forerunner of the turbine is *22X*, James Rumsay's August 26, 1791, grant for an "Improvement of Dr. Barker's Mill."[19] While we don't have a copy of Rumsey's patent, we do have an article in the 1793 *Transactions of the American Philosophical Society*, "Investigation of the Power of Dr. Barker's Mill, as Improved by James Rumsey, with a Description of the Mill" [245] by the American hydraulic mathematician, William Waring, that might have shed some light on Rumsey's invention but didn't.

The conclusion of Professor Warning's mathematics—that turbines were no more efficient than undershot wheels—had the same unfortunate impact on the study of the hydraulic turbine that Parent's naïve model had on the study of the water wheel; viz. to choke off investigation of what would ultimately prove to be a monumental improvement in harnessing the power of flowing water. In reviewing Waring's work a century later, J.A. Drake bemoans its consequence:

> [I]t is stated that the principles of reaction wheels had been fully investigated analytically in examining the merits of Rumsey's improvements on Barker's mill; and the conclusion come to, after a train of reasoning based upon scientific principles, was, that "action and reaction are equal;" that the undershot-wheel is propelled by the action; and Barker's mill by the reaction of the same agent, or momentum; therefore their mechanical effects must be equal.

---

[19]Rumsay was also granted patents *22X, 23X, 24X, 25X, 26X,* and *27X. 24X* is his improvement in Savery's steam engine. *27X* is his much more well-known patent for "propelling boats or vessels." *28X* in the gap is John Fitch's famous patent on "propelling steam boats by steam" while the immediately following patents, *31X, 32X,* and *33X,* are John Stevens, Jr.'s, equally famous inventions for generating steam. The two intervening patents, *29X* and *30X,* went to Nathan Read of Salem, Massachusetts for an improved boiler for steam engines and an improvement in distilling respectively. Steam power was obviously all the rage. As indicated by the italicized patent numbers, none of the original text of these patents survived the 1836 fire.

This conclusion no doubt tended to retard any effort at improvement of
wheels on that principle for a considerable length of time; for it is only,
comparatively speaking, quite recently that reaction water-wheels, of the
form at present in use, have occupied a prominent position before the
public. [74, p. 163]

Table 8.8, abstracted from a table in a history of American turbines published in
the *Transactions of the American Society of Civil Engineers* [223, p. 1302], lists some of
the landmark US patents in the story of the hydraulic turbine.

Grant Date	Number	Inventor
Mar. 19, 1804	535X	Benjamin Tyler, Claremont, NH
Oct. 29, 1829	5,684X	Zebulon and Amasa Parker, Cohocton, OH
Oct. 22, 1830	6,207X	Calvin Wing, Gardiner, ME
May 30, 1838	759	Nelson Johnson, Erwin Center, NY
July 26, 1838	861	Samuel B. Howd, Geneva, NY
Oct. 18, 1839	1,376	Timothy Rose, Windsor, NY
June 27, 1840	1,658	Zebulon and Amasa Parker, Newark, OH
Apr. 30, 1842	2,599	Samuel B. Howd, Arcadia, NY
July 8, 1842	2,708	Reuben Rich, Albion, NY
July 8, 1843	3,158	Whitelaw Stirratt, England
Mar. 26, 1844	3,510	Nelson Johnson, Rathboneville, NY
May 21,1845	4,056	James Leffel, Springfield, OH
May 1, 1847	6,090	Uriah A. Boyden, Boston, MA
June 5, 1847	5,144	Uriah A. Boyden, Boston, MA

Table 8.8: Landmark Hydraulic Turbine Patents

Debates swirling around priority claims often chase their own tail particularly
when they concern an invention with an historic tail as long as the hydraulic turbine's.
Nevertheless, it is generally agreed that Samuel Howd hasn't received the credit he is
due in part, it seems, because he was so reclusive.[20] Howd's inventions demonstrate,
I think, that the motivations and experimental habits of the tinkerer and the savant
are not as different as is often thought. A discussant for the American Society of Civil
Engineers history, Robert E. Horton, observed:

---

[20]This personality trait seems be a characteristic of brilliant antebellum hydraulic engineers. Uriah A.
Boyden, James Emerson, and John Drummer are all described in similar terms.

The average millwright had a sort of superstitious belief in the magical efficacy of centrifugal force, which it was well known augmented the discharge and increased the power of outward flow wheels. In the light of these circumstances, the writer is inspired with a feeling of great respect for the moral courage and creative imagination of Howd who undertook, practically without precedent, and in the face of such well established opposition, to develop and commercialize a turbine water-wheel acting on directly the opposite principle by having the water flow inward instead of outward. [223, p. 1305]

Figure 8.9 shows the drawings from Howd's patents.

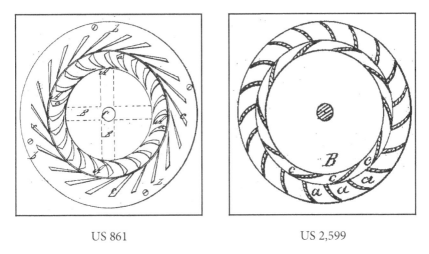

US 861   US 2,599

Figure 8.9: Drawings for Samuel Howd's Patents US 861 and US 2,599

Another reader in experimental philosophy at the Athenæum, Uriah A. Boyden, received five patents for his improvements of the Howd-Fourneyron hydraulic turbine, three of them in 1847 and two more in 1853.[21] Boyden's improvement in the turbine is analogous to James Watt's improvement in the steam engine. Neither Boyden or Watt are credited with the invention of the machine they improved but the improvement of each of them had a game-changing impact on the technical efficiency and commercial success of those machines, Watt for his condenser and Boyden for his diffuser.

Boyden borrowed the *Franklin Journal* only once, on March 20, 1847. Figure 8.10 shows this charge along with all his other charges in 1845 and 1847. The charge of the

---

[21] US 5,068, US 5,090, and US 5,144 in 1847 and US 10,026 and US 10,027 in 1853.

Franklin Journal is for Volume 9.

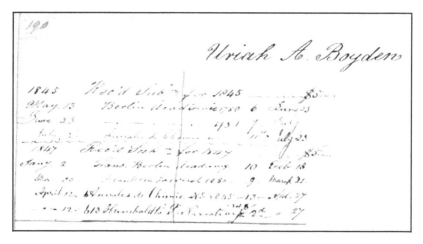

Figure 8.10: Charges of Uriah A. Boyden in 1845 and 1847

Volume 9 of the *Franklin Journal* contains the following notices describing recently issued patents:

Page 33 For an improvement in the Application of Hydraulic Power; Zebulon and Austin Parker, Coshocton county, Ohio, October 19.

Page 161 For an improvement in the Grist Mill, and in Horizontal Spiral Wheels; Alexander Temple, Brookfield, Trumble county, Ohio, December 11.

Page 368 For a Horizontal Water Wheel, called the 'Union Water Wheel;' John R.Wheeler, Pittsford, Monroe county, New York, March 15.

It is conceivable that Boyden was preparing the applications for his 1847 patents when he charged this volume and would as a result be interested in prior art, particularly patented prior art. Zebulon Parker, in particular, was gaining a reputation for sharp elbows in asserting his 1840 patent, US 1,658, "Improvement in Percussion and Reaction Water-Wheels," against infringers.

In commenting on the Parkers' 5,684X patent, the editor of the *Franklin Journal*, Thomas P. Jones, is dismissive of what he calls Barker's mills:[22]

---

[22]Barker's mill is Hero's aeolipile, worked with water rather than steam. In Barker's case, water squirts out of a revolving spout in one direction and pushes the spout around in the other direction. Water wheels

It will be seen from the foregoing, that there is considerable complexity in the apparatus patented; and it may be collected, that the plan proposed consists, mainly, of enclosed wheels, with curved buckets, and certain spiral convolutions in the openings through which the water is to act. It is in fact a modification of the principles of Barker's and the tub wheel....In practice Barker's mill has been abandoned. [145, p. 34]

Jones' dismissal of the Parkers' patent is hard to parse. The principle on which the tub wheel operates is Newton's second law: the impulse force of the percussive force of the water on the paddle. The principle on which Barker's mill[23] operates is Newton's third law: for every action, there is an equal and opposite reaction. Neither of these principles had been abandoned by working hydraulic engineers or any other inventors of the day. Indeed, the hydraulic turbine can reasonably be thought of exactly as Jones describes, the combination of the flow-through property of the tub mill with the reactive property of Barker's mill.

But then Jones was equally dismissive of Howd's patent as he was of the Parkers':

This is another of the so-called reacting water wheels; the claim is to "the application of the water upon the outside of the wheel, and operating upon the principle of reaction, by discharging inwardly, with the spouts or shutes,[sic] giving the water a direction with the motion of the wheel, applied to a reaction wheel as aforesaid."

This wheel does not contain any thing to take it out of the family to which it belongs, and which, unfortunately, is not one of very high standing, although each of the children has been heralded as a prodigy at the time of its first appearance on the stage. [150, p. 23]

These two stinging critiques by Jones are doubly strange because back in 1830 he took a different position when commenting on the Wing's 6207X patent for a re-acting water wheel. Introducing three testimonials to the effectiveness of the Wing wheel, Jones rhapsodizes:

It affords us pleasure to insert the following testimonials of the operation of Mr. Wing's water wheel. We have always thought the employment of the reaction wheel a very wasteful mode of applying water, as it had proved to be so in all the forms under which we have ever known it

---

may be thought to have joined the aeolipile in the dustbin of history but as of this writing the Water Wheel Factory in Franklin, North Carolina, will be happy to build and install a water wheel for you on a stream of your choice.

[23]The original Barker's mill was invented by Johann Segner in around 1760 to test Parent's theory of water wheels, not as a commercial source of power. See Stigler's law of eponymy.

tried. Mr. Wing, however, bids fair to prove that, under the arrangements adopted by him, it will conquer all prejudices, and establish itself in the good opinion of the public, upon the basis of its real merits. [146, pp.89–90]

Maybe the fact that Wing's patent came with testimonials while the Parkers' and Howd's didn't had something to do with Jones' disjointed reviews.

While Boyden designed and supervised the construction a number of hydraulic turbines he made most of the income from his patents by licensing them rather than by practicing them. The authors of the American Society of Civil Engineers history of the hydraulic turbine write:

> [Boyden] accumulated a considerable fortune from the sale of his patents on water-wheels, which, during the later years of his life, enabled him to devote his time to the study of pure science, without consideration of financial return. [223, p. 1251]

But when he did license his patents his elbows were just as sharp as the Parkers'. Boyden was so confident in his turbine mathematics that he contracted to be paid on the basis of how much better his turbine was than the water wheel replaced. The following is reprinted from the *New York Evening Post* in 1857 under the headline "Philosophy in Court"

> This gentleman was concerned in a suit last year, brought by him in the Supreme Court of Massachusetts against the Atlantic Cotton Mills of Lawrence, which was of a very interesting character, but has never, so far as we are aware, come before the public. Mr. Boyden had agreed to make a turbine water-wheel for the Atlantic Mills, which should save or "utilize," as it is termed, seventy-six percent of the water power; if he succeeded in saving that percentage, he was to have $2000; if not, he was to have nothing; and for every one percent above that he was to receive $350. Mr. Boyden went to work and produced a wheel which saved, as he affirmed, *ninety-six; percent.* The labor involved in this result may be imagined from the fact that Mr. Boyden spent more than $5000 in the mere mathematical calculations. The Company had provided no sufficient means of testing the question practically, and as the percentage claimed by Mr. Boyden was altogether unprecedented, they contested the claim.

> The case went into court. No jury on the globe could comprehend the question, and the learned bench also found itself entirely at fault. The case was accordingly referred to three well chosen parties: Judge Joel Parker of Cambridge, Prof. Benjamin Peirce, the mathematician, and James B.

Francis of Lowell, the agent of the united companies of Lowell, in the management of the common water power. Prof. Parker furnished the law, Mr. Francis the practical acquaintance with hydraulics, and Prof. Pierce the mathematical knowledge. That learned geometer had to dive deep and study long before the problem was settled. But settled it was, at last, and in Mr. Boyden's favor, to whom the referees awarded the sum of eighteen thousand seven hundred dollars. Mr. Boyden had previously constructed turbine wheels that utilized respectively the extraordinary amounts of eighty-nine and ninety percent; the last wheel, utilizing ninety-six percent, exceeds anything of the kind that was ever made. The wheel is one hundred and four and three-quarters inches in diameter. [88]

The turbine that Boyden had designed and built for the Atlantic Cotton Mills was not identical to either of the two 102-inch Boyden turbines that became National Historic Mechanical Engineering Landmarks on May 28, 1975, but it was close.[24]

8.3.4   Daniel Treadwell and Spinning Rope

Excellence in engineering is an inscrutable blend of creative art and inventive science. Why and how the cooperation between these two disciplines produces such remarkable results is a recurring topic in the history of technology. *Engineering and the Mind's Eye* by Eugene Ferguson [103] and *Thinking with Objects* by Domenico Meli [175] are two recent explorations of this topic.

The individuals who have achieved this engineering grace and who come readily to mind are by-and-large artists who learned science rather than scientists who learned art: John Whipple and Alexander Parris, to cite two who were readers at the Boston Athenæum. This may be because it is easier to recognize the acquisition of technical knowhow than it is to recognize the acquisition of an artistic sense.

When an inventor is granted multiple patents, they are typically all for inventions in the same domain of technology. Erastus B. Bigelow's many patents are all for rug looms of one sort or another. Athenæum readers Enoch Robinson and Leonard Foster sparkled with new ideas about door and window hardware. Cyrus Alger created a long list of new ways of making things with iron. Joseph Hurd, Jr., envisioned all sorts of new possibilities in stoves and ovens. And nobody knew spinning machines like Paul Moody. Daniel Treadwell is a notable exception to this rule however. Treadwell found inventive opportunities wherever he went.

As so preceptively told by Eugene Ferguson in *Mind's Eye*, a distinguishing characteristic of an individual who thrives at the overlap of science and art is in seeing a

---

[24]...within $2\frac{3}{4}$ inches, one might say.

problem and its solution in the mind's eye. In the telling of his invention of the separate condenser, James Watt[25] famously recalled:

> I had not walked farther than the Golf-house when the whole thing was arranged in my mind. [205, p. 116]

Whether or not these inventions formed in the mind's eye change the course of history as Watt's did, they all have the property that they just fit; they sense a missing part in an existing assembly and bring that missing part into being. Treadwell's iron tail, much like Watt's condenser and Boyden's diffuser, is this kind of invention.

Rope making at the Charlestown Navy Yard was a guild profession. The quality of the rope produced by the navy yard turned on the skills of the men who made it, strand-by-strand and hand-over-hand. As with all guild-controlled manufacturing processes, the cost of the rope was inflated due to the guild's tight control of the supply of rope-making skills. With the intention of decreasing cost and increasing both quality and efficiency, Treadwell set out to automate the process. The machine he invented for the purpose he called Gypsey.[26] [234]

In the early days of the US Patent Office, the fee for filing a patent application was modest. In his history of the early patent office, Kenneth Dobyns provides some details of the fees:

> There was a charge of 50 cents for receiving and filing the petition, and a charge of 10 cents per 100 words for copying the specification to copy sheets. The cost of making out the patent was two dollars, with an additional cost of one dollar for affixing the Great Seal, and finally 20 cents for intermediate services leading to the endorsement of the patent with the date of delivery to the patentee. [73, p. 31]

That the patent application fee included a per word charge undoubtedly served to make the applications less verbose than they needed to be for full and complete disclosure.[27] This fee did not slow the hand of either Treadwell or Bigelow however. While their patents might be thought to be brief by today's standards they are notably long and detailed as compared to others of their day. An exception to this exception is Treadwell's patent 7,997X. It is only two pages long.

---

[25]While renowned for his work with steam engines, Watt's inventive inspirations much like Treadwell's were by no means limited to steam engines.

[26]In his memorial to Treadwell, Morrill Wyman comments on Treadwell's machine: "In perfection and utility, Treadwell's Gypsey ranks with Arkwright's spinning frame; in ingenuity it far excels it; and they stand side by side in the character of their respective products." [257, p. 481]. Treadwell didn't suffer anywhere near the intensity of Luddite attacks that were visited on Arkwright but Morrill does note that "During the whole period he met with determined opposition from the trade of rope makers, was often insulted, and even threatened with violence. [*Ibid.* p. 476]

[27]In many of the X patents one finds a word count penciled in below the application's signature at the end of the application.

Fortunately for our purposes, Treadwell's iron tail patent was described in Volume 18 of the *Franklin Journal*:

> For a new Machine to be used in Rope Making, and which is denominated an "Iron Tail;" Daniel Treadwell, Boston, Massachusetts, February 5.
>
> In the business of rope making, what are called rope tails are used for the purpose of holding back the "top," so that the proper quantity of turn may be given to the lay. In the instrument before us, the strands to be laid are pressed upon by iron bolts, which move in suitable slides upon a flat disk, so that their ends are directed towards the centre of the disk, through which the strands are to pass. The bolts have grooves, or hollows, on their ends, to embrace the strands with sufficient force, which force is regulated by screws, fixed upon bow springs, and bearing against the outer ends of the bolts. The patentee observes, that "various modifications may be made in the different parts of the machine, or instrument; but declares the character of this invention to be comprised in the construction and use of rubbers formed of some solid body, by which they are capable of preserving their own figure, and of constraining the rope over which they pass, to assume the figure defined by them." [147, p. 174]

Figure 8.11 is the drawing in Treadwell's iron tail patent.

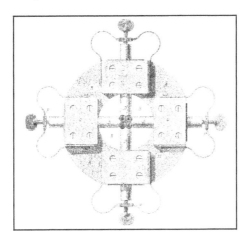

Figure 8.11: Drawing of Daniel Treadwell's Iron Tail, Patent 7,997X

In addition to selling his Gypsey machine to the Department of the Navy for use in the ropewalk at the Charlestown Navy Yard, in 1833 Treadwell in league with two other of our readers, Francis Calley Gray and Horace Gray, formed a joint stock company named simply the Spinning Company to manufacture rope on the Mill-dam. The

Spinning Company evolved into the Boston Hemp Manufacturing Company which made cordage in Boston until 1858.

The tension between Treadwell and the rope-making guild came to a head in 1840 when the Board of Navy Commissioners held a rope-off between man and machine. As reported by Wyman:

> The result was, that the machine-spun rope sustained before it broke a weight of 1,469 pounds, while the hand-spun rope of the same size broke with a weight of 1,278 pounds. By other and careful experiments made at the same time, it was shown that the cost of spinning one ton of hemp by hand was $29.25, and by the machine $14.13; that is, half the cost of that by hand. [258, p. 377]

The strength of rope is determined by the strength of its weakest twist. While the rope spun by the guild may have been stronger at points than rope made by Treadwell's Gypsey with its iron tail, Gypsey's rope was of uniform strength throughout, a strength greater than the guild's weakest link.

8.3.5   George M. Dexter and Drying Gunpowder

George M. Dexter was almost the architect of the Boston Athenæum's current home, twice. He won the competition to build the Tremont Street successor to the Pearl Street building but the proposed Tremont Street location was abandoned in favor of the current Beacon Street location. Dexter submitted a second design for this new location but lost out to Edward Clarke Cabot for a number of reasons not all of them having to do with engineering or artistic considerations. The entire story is told in Catharina Slautterback's *Designing the Boston Athenæum: 10½ at 150*.

Like one of his mentors, Alexander Parris, Dexter was as much an engineer as he was an architect. He trained with Loammi Baldwin as did Parris and worked with Parris on some of the Charlestown Navy Yard projects. His engineering talents were sufficiently highly regarded that if the Athenæum's current location could be said to have a architectural engineer it would be Dexter. He was put in charge of the construction of the building at 10½ Beacon Street while Cabot "attended principally to the office business and drawing." [217, pp. 32–37]

Dexter was granted US 1,875 on December 1, 1840. The full title of the patent is

Improvement in the Mode of Heating Buildings by Means of an Apparatus Consisting of Tubes for the Circulation of Hot Water Arranged in an
Air-Chamber Adapted to the Same.

Figure 8.12 is the sole drawing in Dexter's 3-page patent application.

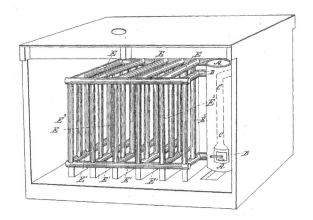

Figure 8.12: Drawing for George M. Dexter's Patent US 1,875

The intended utility of Dexter's invention is to heat air to be distributed to living
spaces but more particularly for "drying-rooms of laundries, bleacheries, manufactories of paper, gunpowder, and many other articles."[28] The advantage of his heater is
that it employs hot water rather than steam and as a result could be promoted as being
safer.

The air chamber in Figure 8.12 is nominally ten to twelve feet in height. In the
lower right corner is a firebox marked **B** at the base of a vertical cylinder **AA** which
contains water. The fire heats the flue **CC** which in turn heats the water in the cylinder.
As the water closer to the bottom of the cylinder and nearer to the firebox will be hotter
than the water at the top of the cylinder it will rise pushing the water at the top of the
cylinder into the pipe **D** and hence into the pipe network: across the pipes E, down
pipes E^2 and, finally, across pipes E' back to the cylinder. In the patent application,
Dexter describes how his invention works:

D is the pipe leading from the upper part of the boiler to the frame-work
of the tubes **EEEE** and **F** is the return-tube leading into the lower part
of the heating-vessel and reconveying the water into it after it has passed
through the frame-work of tubes.

---

[28]There is no mention of athenaeums.

In *Engineering and the Mind's Eye* [103] Eugene Ferguson gives some examples of mechanical drawings done by copyists without an understanding of the machine whose drawing they were copying. The consequence was at times a drawing of a machine that couldn't work at all nevermind in the way the inventor intended.

The attentive reader will have noticed that there is no tube marked **F** in the drawing in Figure 8.12. It is obvious which pipe should be marked **F** but it is curious that Dexter, skilled as he was in architectural drawings, would have made this mistake. In fact, it is likely he didn't.

Kenneth Dobyns, a patent attorney and the author of a book on the early history of the United States patent office [73], notes:

> The drawings as published were not drawn by the applicant or his attorneys. In 1840, they had not started printing patents, so drawings as submitted by applicants did not have to be black and white line drawings. I have read that Robert Fulton submitted a bound book of beautiful water color drawings (of course the book was burned in 1836). When they finally started printing drawings in the 1860s and 1870s, the patent office draftsman turned out large numbers of printable black and white drawings for use in printing old patents. Perhaps in this case, the patent office draftsman missed the letter F on the original drawings.[29]

8.3.6   Samuel Goodrich and Fixing Plates

Mark Twain was granted three patents:

- US 121,992 granted on December 19, 1871 for an Improvement in Adjustable and Detachable Straps for Garments
- US 140,245 granted on June 24, 1873, for Improvement in Scrap-Books
- US 324,535 granted on August 18, 1885, for a Game Apparatus

The pages of Twain's improved scrapbook, US 140,245, were coated with paste so one need only moisten the page to archive that ticket stub or dance card.[30] The scrapbook is said to have sold 25,000 copies and earned Twain $50,000 compared to $200,000 for all of his other books combined which, he observed, "was well enough for a book that did not contain a single word that critics could praise or condemn."

Samuel Goodrich, about whom we will learn more in the following chapter, was a Boston publisher and, like Twain, the author of children's literature. Goodrich was granted patent *5,243X* on October 11, 1828, for "an improved mode of constructing

---

[29]Personal communication.

[30]Twain touted additional advantages for his scrapbook including the prevention of swearing and the curing of "inflammatory rheumatism complicated with St. Vitus's dance." These are not however listed in the patent's claims.

stereotype blocks." The patent document is not among those which have been recovered but it was briefly described in the December 1828 issue of the *Franklin Journal* as follows:

> This is a very simple and neat contrivance, for fixing stereotype plates upon a wooden block. A strip of brass is firmly screwed on one edge of the block, and projects, in two places, above its side, so as to form a lip, to receive one edge of the stereotype plate. A notch is cut on the opposite side of the block, to receive, and allow play to a moveable lip of brass, which is to confine the other edge of the plate. This moveable lip is perforated with three holes in a row, as in the margin. The two outer holes have wires soldered into them, which project out about two inches, and slip, neatly, into corresponding holes in the edge of the block. From this same edge projects a screw, which passes through the middle hole; upon this screw a nut is fitted, and is turned, first by the fingers, and then by a small wrench, so as to cause the projecting lip to embrace the plate firmly. A brass plate, the whole length of the block, is screwed upon its edge, so as to cover the notch of the moveable lip. This plate is hollowed at its upper edge, opposite the nut, to allow it to be turned with facility.[144, p. 400]

For I could not but consider, that being yet but very young, not only in years, but, what is much worse, in experience, I have yet much more need to learn, than ability to teach.

A Proëmical Essay
Robert Boyle

Chapter 9

Farmers and Children

9.1 Daniel Treadwell and the Committee on Useful Inventions

Daniel Treadwell was a member of the Committee on Useful Inventions of the Massachusetts Society for Promoting Agriculture.[1] The committee posted rewards for "Experiments, Discoveries, and Inventions" and awarded premiums at the annual Brighton Cattle Show. Initially the premiums were modest, in the range of $10 to $20, but they could climb higher for challenges that were regarded as being particularly difficult or of pressing concern. The very first premium announced in 1801 was to be awarded...

> ...to the person who shall discover an effectual and cheap method of destroying the Canker-worm, and give evidence thereof, to the satisfaction of the Trustees, on or before the 1st day of October, 1803, a premium of one hundred dollars, or the Society's gold medal. [238, p. 30]

By 1857, a premium of $1,000 was offered...

> ...for the best plantation of trees, of any kind commonly used for, and adapted to, ship-building, grown from seed planted for the purpose, or otherwise, on not less than five acres of land, one white oak at least to be planted to every twenty square yards. [238, p. 149]

---

[1] One of the founding fathers of the Boston Athenæum, Rev. Joseph Stevens Buckminster, was a trustee of the society. Incorporated in 1792, it is still going strong today. Readers Paul Moody, Cyrus Alger, and John Heard, Jr., were all members of the committee on inventions which for many years was chaired by Josiah Quincy.

Entries for premiums were by no means restricted to those on the society's published list but could be "for any newly invented Agricultural Implement, or Machine, superior to any designed for the same use." In 1827, for example, Treadwell's committee awarded John and Horace M. Pool a $15 premium for…

> …several Geometrical Protractors, of a new construction, by which lines may be drawn with great facility and at any required angle to the side of the tablet or drawing board, which in this case forms a base line; it is therefore a most convenient instrument in forming plans of surveys, an operation of primary importance to the farmer, considering the simplicity of the instrument, and the ease with which it may be applied. [81, p. 237]

Nor were premiums offered only for agricultural innovations or for solving agricultural problems of the day. Premiums were also offered for conducting experiments. In the same issue of the *Massachusetts Agricultural Repository and Journal* that contained the announcement of the award to the Pools for their geometric protractors there is a list of six premiums for "Agricultural Experiments" including a premium of $20 that would be awarded…

> …to the person who shall make the experiment of turning in Green Crops as a manure, on a tract not less than one acre, and prove its utility and cheapness, giving a particular account of the process and its result. [237]

In the early 1820s a topic of pressing concern to Massachusetts farmers was the Rose Bug. The Massachusetts Agricultural Society had posted a premium for ideas about controlling the bug. In an article in Volume 9 of the *Massachusetts Agricultural Journal* the trustees of the society bemoan the fact that nobody had as yet submitted a claim for the premium:

> The Trustees of the Massachusetts Agricultural Society having received from various parts of the State, in the summer of 1825, accounts of the extensive devastation and injury produced by this insect, were induced to offer a premium to any person who should produce an essay on its natural history, and point out any probable means of checking their progress. No such essay has appeared, to our deep regret. [105, pp. 143-44]

The trustees' entreaty did not go unheard. The very next volume of the journal happily reports:

> The following essay upon the natural history of the Rose Bug, was prepared by Dr. T. W. Harris of Milton, with a view to the premium offered by the Massachusetts Society, for promoting agriculture, for the best essay on this subject; but the professional, cautious of the amiable and learned

writer having prevented the completion of it within the period fixed by the Trustees, he had resolved to suppress it. The progress which he had made, having come to the knowledge of the President of that Society, he urged Dr. Harris to lay it before the Trustees in its present state, and they were pleased to award the Society's premium to the author. We think the readers of this journal will feel obliged to us for inserting it.

The title of the essay was "Minutes Towards a History of some American Species of Melolonthæ Particularly Injurious to Vegetation." [126] It was written by Thaddeus William Harris, a reader in experimental philosophy at the Boston Athenæum.

9.2   Purposeful Reading for Farmers

Thaddeus William Harris[2] and his father, Rev. Dr. Thaddeus Mason Harris, were both Harvard librarians and were both avid naturalists.[3] Harris' report was acclaimed when it was first published in 1841 and is still respected in entomological circles today. Before getting to Harris' report, however, a little background on how it came to be written is perhaps in order.

9.2.1   Massachusetts Geological, Zoological, and Botanical Surveys

On June 5, 1830, the Massachusetts legislature called for a geological survey of the state:

> Resolved, That his Excellency the Governor, by and with the advice of the council, be, and he is hereby authorized to appoint some suitable person, to make a geological examination of the Commonwealth, in connection with the general survey, in order that the same may be inserted on the map which may be published.

The suitable person called to the task Edward Hitchcock, Professor of Chemistry and Natural History at Amherst College. Hitchcock delivered his 692-page report a little over three years later. The full title of the report is:

> Report on the Geology, Mineralogy, Botany, and Zoology of Massachusetts Made and Published by Order of the Government of that State: in Four Parts: Part I. Economical Geology. Part II. Topographical Geology. Part III. Scientific Geology. Part IV. Catalogues of Animals and Plants. With a Descriptive List of the Specimens of Rocks and Minerals Collected for the Government.

---

[2] Henceforth, 'Harris' standing alone will refer to Thaddeus William Harris.

[3] The father is the subject of Nathaniel Hawthorne's short story "The Ghost of Dr. Harris." The son was the inventor, if anyone can be said to be, of the library card catalog. [93, p. 129–132]

The first three parts on geology were written by Hitchcock himself and comprise the initial 541 pages of the report. The catalogs of animals and plants comprising Part IV were assigned to recognized experts in each field. Their reports are listed in Table 9.1 along with the number of pages in their contribution.

Field	Author	Pages
Mammals	Edward Hitchcock	2
Birds	Ebenezer Emmons	7
Reptiles	David S.C.H. Smith	1
Fishes	Jerome V.C. Smith	2
Shells	Thomas A. Greene	4
"	John Milton Earle	3
"	Col. Joseph G. Totten	4
Crustacea	Augustus A. Gould	2
Spiders	N.M. Hentz	1
Insects	Thaddeus William Harris	30
Radiata	Augustus A. Gould	1

Table 9.1: Zoological Catalogs in the 1833 Hitchcock Report

Only a geology survey was requested in the original charge from the legislature so it isn't clear why catalogs of animals and plants were tacked onto the final report. Judging from the page counts, all save Harris' insect catalog were cursory to say the least.

A second edition of the Hitchcock report was published in 1835 [135] and from this was extracted a 142-page tract, *Catalogues of the Animals and Plants of Massachusetts with a Copious Index.* The page counts for the animal catalogs aren't noticeably different than the counts in Table 9.1 except that Harris' insect catalog had grown to 50 pages.

There surveys of the flora and fauna of Massachusetts rested until April 12, 1837, when the legislature called for another survey:

> Resolved, That his Excellency the Governor, with the advice and consent of the Council, is hereby authorized and requested to appoint some suitable person, or persons, to make a further and thorough Geological, Mineralogical, Botanical, and Zoological Survey of this Commonwealth, under his direction, particularly in reference to the discovery of Coal, Marl, and Ores, and an analysis of the various soils of the State, relative to an

Agricultural benefit. And he is hereby authorized to draw his warrant, from time to time, upon the treasurer of this Commonwealth, for any sum not exceeding two thousand five hundred dollars, for the foregoing purposes.

Hitchcock's report had given extensive consideration to value that might be excavated from Massachusetts' geology. Now the legislature turned its attention to extracting value from the state's plants and animals.

One of our readers in experimental philosophy, George B. Emerson, was the chairman of the committee that oversaw this second effort. In accordance with the legislature's resolution, Emerson delivered his report on August 13, 1839, but he begins by noting that the schedule set forth in the resolution proved insufficient:

> It was immediately seen that no final reports, that would be satisfactory, in this view of the work, could be made within the limits of a single year. Partial reports were therefore made by the five Commissioners on the Zoological departments, which, with a letter from the Chairman, were ordered to be printed; and leave was asked and obtained, by all the Commissioners, to defer their reports for another year. [119, p. vii]

Table 9.2 lists each of the surveys ordered by the legislature as well as the year in which they were finally published.

Year	Author	Title
1839	D. Humphreys Storer	*Reports of the Fishes, Reptiles and Birds of Massachusetts*
1839	Wm. B.O. Peabody	*Reports of Birds of Massachusetts*
1839	D. Humphreys Storer	*Reptiles of Massachusetts*
1840	Chester Dewey	*Reports of the Herbaceous Plants and on the Quadrupeds of Massachusetts*
1841	Thaddeus William Harris	*A Report on the Insects of Massachusetts, Injurious to Vegetation*
1841	Augustus A. Gould	*Report on the Invertebrata of Massachusetts, comprising the Mollusca, Crustacea, Annelida, and Radiata*
1846	George B. Emerson	*A Report on the Trees and Shrubs Growing Naturally in the Forests of Massachusetts*

Table 9.2: Biological and Zoological Surveys of Massachusetts

Table 9.3 lists the number of titles, readers, and charges for survey categories other than entomology—namely, zoology, ornithology, and conchology—found in Emerson's report. Figure 9.1 is the cumulative charge plot of each of these three. In the figure the solid curve is zoology, the heavy dashed line is ornithology, and the light dashed line, mostly between the later two, is conchology.

Subject	Titles	Readers	Charges	Interest
Zoology	40	107	323	37
Ornithology	18	67	122	14
Conchology	12	44	114	13

Table 9.3: Titles, Readers, and Charges in Zoology, Ornithology, & Conchology

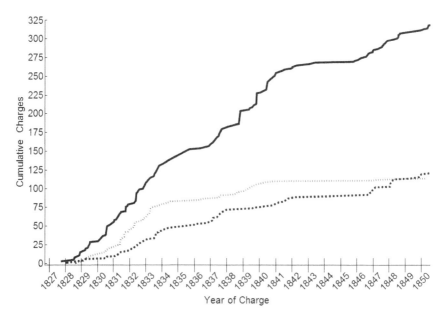

Figure 9.1: Plot of Charges in Zoology, Ornithology, and Conchology

9.2.2   Readers in Entomology

Eighty-nine Athenæum readers generated 221 charges against twenty-four titles in entomology. Figure 9.2 is the cumulative charge curve and Table 9.4 contains the details of the single period of elevated interest.

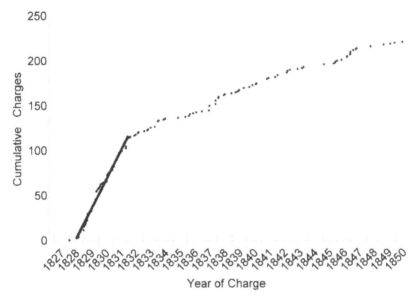

Figure 9.2: Plot of Charges in Entomology

Period	Start	End	Readers	Charges	Interest
	Jan. 1827	Dec. 1850	90	222	25
A	Aug. 1828	Dec. 1831	41	114	89

Table 9.4: Periods of Elevated Interest in Entomology

The sudden fall in the interest in entomology at the end of 1831 is curious. The interest value for 1832 through 1850, the right-hand segment of the cumulative charge curve, is 16 as compared to 89 for the left-hand segment covering roughly 1830 and 1832. The number of charges, number of readers, and number of titles charged in the two periods are all about the same. Sorting out the top readers and top titles in the two periods provides some insight. These are listed in Tables 9.5 and 9.6 respectively.

Kirby and Spencer's *Introduction to Entomology* held onto first place but second and third place both turned over. Second place changed from Huber's *Natural History of Ants* to Huber's *Natural History of Bees* and third place changed from Say's *American Entomology* to Harris' *Insects of Massachusetts*. The precipitous shift of interest from ants to bees is a bit of a puzzle but it is no surprise that a book about injurious local bugs by a prominent member of the Athenæum and published in 1841 would attract the attention of Athenæum readers.

	1827–1831		1832–1850	
Short Title	**Readers**	**Charges**	**Readers**	**Charges**
Introduction to Entomology	21	40	15	21
Natural History of Bees	4	4	14	17
Natural History of Ants	12	13	7	7
Insects on Fruit Trees	5	5	6	9
Insects of Massachusetts	0	0	14	14
Dialogues on Entomology	8	9	4	4
American Entomology	7	11	1	2

Table 9.5: Popular Titles in Entomology by Period

1827–1831		1832–1850	
Reader	**Charges**	**Reader**	**Charges**
Parsons, Theophilus	13	Harris, Thaddeus M.	6
Codman, Henry	9	Weston, Alden	5
Osgood, David	8	Harris, Thaddeus W.	5
Capen, Nahum	7	Bowditch, N.I.	4
Quincy, Josiah, Jr.	6	Burley, Susan	4

Table 9.6: Top Readers in Entomology by Period

9.2.3 Thaddeus William Harris' *Insects of Massachusetts*

As as example of the study of entomology for practical purposes, consider:

> Few persons while indulging in the luxury of early green pease [*sic*] are aware how many insects they unconsciously swallow. When the pods are carefully examined, small, discolored spots may be seen within them, each one corresponding to a similar spot on the opposite pea. If this spot in the pea be opened, a minute whitish grub, destitute of feet, will be found therein. [127, p.55]

This observation on life forms in peas is contained in Harris' *A Report on the Insects of Massachusetts, Injurious to Vegetation*. As the topic of the report was the damage that insects cause to Massachusetts flora, Harris goes on at length to describe the impact of the whitish grub on the pea.[4]

Harris' biography is well-told in *Thaddeus William Harris 1795–1856: Nature, Science, and Society in the Life of an American Naturalist* [93] by Clark A. Elliott. A thread that runs through Elliott's telling—a thread that might seem surprising given Harris' position as Harvard's librarian—is the difficulties he faced gaining access to books on entomology, particularly those of British and European authors.[5] Elliott recalls:

> At this time [1826], Harris was preparing a list of entomological books for [John] Lowell, who had promised to procure certain works for the Boston Athenæum library where Harris already had access to other books of interest to him. The Athenæum was an important resource for him but he recognized that other subject interests there competed heavily with natural history. [93, p. 56–57]

Table 9.7 is a list of the entomology titles charged by father and son at the Athenæum. It is not known if any of these were on Harris' list.[6]

---

[4]There is no subsequent mention of any impact on humans due to consumption of the grub.

[5]In a letter to the Harvard Corporation, Harris noted that "[T]he Boston Athenæum [has] many more works on natural history that does the University library." [93, p. 57]

[6]In addition to borrowing books, Thaddeus William Harris lectured on entomology at the Athenæum [93, p. 182] and his son, Thaddeus William, Jr., was an assistant librarian at the Athenæum before his untimely death at the age of twenty-eight. [93, p. 73]

Author	Short Title	Charges
DeGeer	*Mémoires pour Servir à l'Histoire des Insectes*	3
Stoll	*Aanhangsel van Cramers Uitlandsche*	2
Crammer	*Uitlandsche Kapellen*	2
Major	*Insects on Fruit Trees*	2
Diderot	*Encyclopedie Methodique [Insects]*	2
Donovan	*Natural History of British Insects*	1

Table 9.7: Entomology Titles Charged by T.M. and T.W. Harris

Harris lectured in entomology at Harvard from time to time but never realized his life-long desire to obtain a professorship. The Fisher Professorship in Natural History opened up in 1842 but it was ultimately awarded to Asa Gray. Harris may have been passed over because he was doing too good a job at the library but it is more likely to have been because his publications appeared in both scientific and practical journals. Elliott notes:

> Harris's scientific work was never simply the pursuit of knowledge for its own sake. It *was*[7] that, but his genuine interest in popularization as a strategy for promoting his subject, and his life-long interest in the applications of entomological knowledge for practical purposes, preclude easy characterization of his life as a scientist. [93, p. 144].

Harris' list of publications, found at the end of Thomas Wentworth Higginson's memorial [133], contains 114 entries. Most of these are short notes addressed to the agricultural community and appearing in publications such as the *New England Farmer* and the *Massachusetts Ploughman*. But the list also includes longer articles in more scholarly journals such as the *American Journal of Science*, the *Annals of Science*, *Boston Journal of Natural History*, and the *Proceedings of the Boston Society of Natural History*. That he was by necessity an avocational entomologist, however accomplished, and that he published in the popular as well as the scholarly literature must have weighed against him when the short list for the Fisher Professorship was drawn up.

Whether he would have wished it or not, his report, *Insects of Massachusetts*, is how he is remembered. The work is still regarded today as a "classic in economic entomology" and an "unrivalled introduction to the subject." [93, p. 215] The report saw three editions and five impressions all of which are in the Athenæum's collection. These are listed in Table 9.8.

---

[7] Emphasis Elliott's.

Edition	Year	Citations
First	1841	[127]
First	1842	[128]
Second	1852	[129]
Third	1862	[130]
Third	1862	[131]

Table 9.8: Editions and Impressions of Harris' *Insects of Massachusetts*

In his cover letter to George B. Emerson, the chairman of the committee that commissioned his report, Harris records why he writes for trade publications in addition to more scholarly journals:

> I have felt it my duty, in treating the subject assigned to me, to endeavour to make it useful and acceptable to those persons whose honorable employment is the cultivation of the soil. Some knowledge of the classification of insects and of the scientific details of entomology seems to be necessary to the farmer, to enable him to distinguish his friends from his enemies of the insect race. [127, p. v]

Harris describes eloquently, if not in so many words, the bridge between the academy and the farm that he chose to build and travel.

9.3 Purposeful Reading for Children

Richard Green Parker was the principal of Boston's Johnson Grammar School for many years. His book

> *The Boston School Compendium of Natural and Experimental Philosophy, Embracing the Elementary Principles of Mechanics, Hydrostatics, Hydraulics, Pneumatics, Acoustics, Pyronomics, Optics, Electricity, Galvanism, Magnetism, Electro-Magnetism, Magneto-Electricity, and Astronomy, with a Description of the Steam and Locomotive Engines* [187]

was adopted as a textbook by the Boston School Committee in 1836 and used in the Boston school system for many years thereafter.

At the same time the School Committee selected Parker's text they also adopted on a trial basis a short list of philosophical apparatus to be used in teaching from the book. The trial must have been regarded as successful since a decade on the School

Committee adopted a "Schedule of Philosophical Apparatus for the Boston Grammar Schools." The schedule enumerates over sixty items including Miser's plate for shocks, copper chamber for condensed air fountain, brass Magdeburg hemispheres, dissected eyeball (listed under astronomy), and an improved school orrery.

In the advertisement to his book, Parker enumerates some of the works he consulted in preparing his textbook. As might be expected, his list includes books on experimental philosophy found in the Athenæum's books borrowed registers including the following:

- Arnott's *Elements of Physics*
- Bigelow's *Elements of Technology*
- Herschel's *Treatise on Astronomy*
- Lardner's *The Steam Engine*
- Silliman's *American Journal of Science*
- Singer's *Elements of Electricity and Electro-Chemistry*
- Turner's *Elements of Chemistry*
- Wilkins's *Elements of Astronomy*

As noted in Chapter 1, the study of experimental philosophy in antebellum Boston was much more than an idle pastime of the learned. It was a way of understanding and putting to use the rapidly-expanding body of scientific and technical information. The adoption of Parker's textbook along with its associated schedule of philosophical apparatus by the Boston School Committee demonstrates that experimental philosophy was taught as such in the Boston school system.

But experimental essays can also be found in the literature read to children even before they went off to grammar school. Consider the following passage:

> You may illustrate this in a simple manner by fastening two stones of equal weight, one at each end of a string, and throwing them to a distance; in their course you will see them revolve in circles about each other, and a common centre, agreeing with the centre of the string. If you now try the experiment with two stones of unequal weight, you will find this common centre in a point of the string nearest the heavier stone; the lighter one being more powerfully acted on, describes larger circles, while the other remains comparatively at rest. [115, p. 37]

This experiment demonstrating the motion of a planetary system is taken from *Peter Parley's Illustrations of Astronomy* by Samuel Goodrich. Goodrich read in experimental philosophy at the Boston Athenæum in order to write about and according to the ways of experimental philosophy for children.[8] The texts of Goodrich as well as

---

[8]Goodrich's book *Peter Parley's Universal History* is listed by the Library of Congress as one of the "88 Books that Shaped America."

another author of children's literature reading at the Athenæum, Jacob Abbott, share a common pedagogical framework: learning through experimentation and discovery. They wrote in as much of the spirit of Montessori and Pestalozzi as did Colburn and Sherwin. Abbott called it Rollo's Philosophy. Today we might call it critical thinking.

When it comes to literary style, however, Goodrich and Abbott are starkly different. Goodrich's books are straight-up lectures flavored with history and mythology to help the medicine go down. Peter Parley, the main character in Goodrich's books, is a kindly teacher, talking to a small group of nameless and faceless children that includes the reader.[9] Abbott's books by contrast are conversations among children and adults all of whom take on distinct personalities and become friends of the reader as the story unfolds. Jacob Abbott's hero, Rollo, is the inquisitive child explorer, peppering the characters in Abbott's stories with a never-ending stream of questions.

9.3.1   Samuel Goodrich and Peter Parley's Astronomy

Hurrah for Peter Parley
Hurrah for Daniel Boone
Three cheers sir, for the gentleman
Who first observed the moon—

Emily Dickenson
"Sic Transit Gloria Mundi"

Samuel Goodrich revolutionized children's literature in much the same way and for many of the same reasons that Warren Colburn revolutionized the teaching of primary school mathematics. Just as Colburn found the arithmetic texts of the day mind-numbing, so Goodrich found that...

> ...much of the vice and crime in the world are to be imputed these atrocious books put into the hands of children, and bringing them down, with more or less efficiency, to their own debased moral standard. [116, v.1 p. 169]

In Goodrich's view, children will learn most readily by means of curiosity-driven discovery and delight aided by educational illustrations. In his *Recollections of a Lifetime* he recalls:

> I came to the conclusion that in feeding the mind of children with facts, with truth, and with objective truth, we follow the evident philosophy of nature and providence, inasmuch as these had created all children to be ardent lovers of things they could see and hear and feel and know. Thus

---

[9]...much like Sesame Street's Big Bird.

I sought to teach them history and biography and geography, and all in
the way in which nature would teach them, that is by a large use of the
senses, and especially by the eye—the master organ of the body as well as
the soul. [116, v.2 p. 308]

The 1837 up-tick in astronomy charges back in Figure 7.5 is due entirely to
Goodrich. There are three books on astronomy in Goodrich's Peter Parley Series:
Tales about the Sun, Moon, and Stars in 1831 [113], *Peter Parley's Dictionary of Astronomy* in 1836 [114], and *Peter Parley's Illustrations of Astronomy* in 1840 [115]. Table 9.9
lists Goodrich's astronomy charges along with the date of each charge during this period. It is of passing interest that he also borrowed Taylor's *Logarithms* on May 27,
1836, in the middle of his astronomy borrowing spree. I find no mention of logarithms
in any of Peter Parley astronomy books so perhaps Goodrich was using logarithms to
work out some of the problems in the astronomy books he borrowed.

Author	Short Title	Charge Date
Vince	*Complete System of Astronomy*	Mar. 28, 1835
Vince	*Complete System of Astronomy*	Apr. 23, 1835
Ferguson	*Easy Introduction to Astronomy*	Feb. 19, 1836
Francoeur	*Uranographics*	Feb. 19, 1836
Vince	*Complete System of Astronomy*	Feb. 27, 1836
Delambre	*Histoire de l'Astronomie Moderne*	Mar. 21, 1836
Ferguson	*Easy Introduction to Astronomy*	Mar. 26, 1836
Biot	*Traité Éleméntaire d'Astronomie*	Apr. 9, 1836
Biot	*Traité Eleméntaire d'Astronomie*	May 28, 1836
Bonnycastle	*Introduction to Astronomy*	June 11, 1836
Brinkley	*Elements of Astronomy*	Aug. 27, 1836
Herschel	*A Treatise on Astronomy*[10]	Oct. 6, 1836
Ferguson	*Easy Introduction to Astronomy*	Oct. 6, 1836
Delambre	*Histoire de l'Astronomie Moderne*	Oct. 6, 1836
Ferguson	*Easy Introduction to Astronomy*	Nov. 5, 1836

Table 9.9: Goodrich's Charges in Astronomy

---

[10]...as Volume 43 of *Lardner's Cabinet Cyclopædia*.

Long before Goodrich published his astronomy books he authored an arithmetic. Writing in his *Recollections* about his experiences as publisher and printer in Hartford, Connecticut, in 1819, he recalls:

> About the same period I turned my attention to books for education and books for children, being strongly impressed with the idea that there was here a large field for improvement. I wrote, myself, a small arithmetic, and half a dozen toy-books, and published them, though I have never before confessed their authorship. [116, v. 2 p. 110]

The Appendix of the second volume of *Recollections* contains a *List of Works of which S. G. Goodrich is the Editor or Author*. The list includes *Parley's Arithmetic* published in 1833 but does not mention any earlier arithmetic texts. Regarding his "small arithmetic" Karpinski shows two early arithmetic texts by Goodrich:

1. [GOODRICH, SAMUEL G.] The child's arithmetic. Being an easy and cheap introduction to Daboll's, Pike's, White's, and other arithmetics; designed to render both teaching and learning at once simple and interesting. Hartford, Samuel G. Goodrich, 118, 4 p.l., [7]-69. [151, p. 217]

2. [GOODRICH, SAMUEL G.; *pseudonym*, PETER PARLEY]: The youth's arithmetic; being a plain and easy method of teaching the practical use of numbers. For the use of schools. Hartford, S.G. Goodrich, printed by Lincoln and Stone, 1819, 216 p., 1 l. or errata. [151, p. 228]

   Editions:
   1820. Second edition. Same publisher, 188 p.
   1825. Hartford, Silas Andrus, stereotyped by A. Chandler, 180 p.

Goodrich doesn't mention Pestalozzi by name in his memoir, but *Peter Parley's Method of Teaching Arithmetic* is listed as among "Formalized Pestalozzian Arithmetics" in Monroe's exhaustive review of early American arithmetic texts. [178, p. 103]

Getting back to astronomy, *Peter Parley's Dictionary of Astronomy* and *Peter Parley's Illustrations of Astronomy* are both glossaries of astronomical vocabulary rather than rambling Peter Parley tales. For example, in the *Dictionary of Astronomy* we find:

> IMMERSION: denotes the beginning of an eclipse, an occultation, or a transit; or the instant a celestial body begins to disappear by the interposition of another, or by entering a shadow. [114, p. 70]

Peter Parley is at his story-telling finest in *Tales about the Sun, Moon, and Stars*. Long before NASA, the gouty teller of tales conjured up a picture of the earth as seen from outer space in the imaginations of his little readers.

Figure 9.3: Peter Parley's View of the Earth from Outer Space

Judging from the footnotes and references[11] in his astronomy books and particularly his expressed admiration for the works of William Herschel, to Wendell's point of whether or not borrowers actually read what they charged, it is safe to say that Goodrich did indeed read the astronomy texts that he borrowed.

9.3.2   Jacob Abbott and Rollo's Philosophy

The second prodigious author of children's books who read at the Boston Athenæum was Jacob Abbott. Abbott's books lean more toward history and moral teaching than do Goodrich's but experimental philosophy was evident in all of the science books in his oeuvre. Some of these are listed in Table 9.10

---

[11]Yes, footnotes and citations to the current scientific literature in children's books.

Year	Short Title
1839	*Rollo's Experiments*
1841	*Rollo's Philosophy, part 1: Water*
1841	*Rollo's Philosophy, part 2: Air*
1842	*Rollo's Philosophy, part 3: Fire*
1842	*Rollo's Philosophy, part 4: Sky*
1871	*Science for the Young, 1: Heat*
1871	*Science for the Young, 2: Light*
1871	*Science for the Young, 3: Land and Water*
1871	*Science for the Young, 4: Force*

Table 9.10: Jacob Abbott's Books in Experimental Philosophy

Abbott published two arithmetic texts with his brother, Charles:

- *The Mount Vernon Arithmetic. Part I. Elementary*

- *Abbotts' Addition Columns, for Teaching the Art of Adding with Facility and Correctness*

and one on his own:[12]

- *The Mount Vernon Arithmetic. Part II. Vulgar and Decimal Fractions.*

I find no connection to Abbott's books in experimental philosophy or arithmetic in his scant entries in the books borrowed registers.

This is not, however, the case when it comes to his history series. The books he borrowed while he was writing *Mary Queen of Scots, History of King Charles II of England, Queen Elizabeth*, and *Cyrus the Great and Romulus* and about a dozen others of the same genre are listed in Table 9.11.

---

[12]See [151, pp. 482–83] and [151, pp. 494–95].

Date	Author	Short Title	Volume
Mar. 29, 1848	Robertson	*Scotland*	v.2
Apr. 1, 1848	Kennett	*England*	v.1, 2
Apr. 7, 1848	Macauly	*England*	v.4
Apr. 7, 1848	Williams	*Alexander the Great, Family Library*	v.3
Apr. 7, 1848	Barrow	*Peter the Great, Family Library*	v.35
Aug. 26, 1848	Aikens	*Elizabeth*	v.1, 2
Oct. 17, 1848	James	*Louis XIV*	v.1
Jan. 9, 1849	Tyrrell	*History of England*	v.1
Jan. 9, 1849	Pike	*Memoirs of Court of Louis 14*	v.1
Nov. 1, 1849	Milner	*Life & Times of the Rev. Isaac Watts*	
Mar. 13, 1850	Thomsons	*Sutonius*	
Mar. 13, 1850	Rollins	*Ancient History*	v.13
Oct. 2, 1850	Duncan	*Caesar*	v.13
Oct. 3, 1850	Rollins	*Ancient History*	v.9
Oct. 3, 1850	Rollins	*Ancient History*	v.8
Oct. 9, 1850	Bundy	*Roman History*	
Oct. 11, 1850	Rollins	*Ancient History*	v.11
Oct. 11, 1850	Hook	*Roman History*	v.1

Table 9.11: Jacob Abbott's Charges in History

Where Goodwin's main character, Peter Parley, is a whimsical professor, Abbott's main character, Rollo, is the learner rather than the learned. To say Rollo was inquisitive would be like calling Robert Boyle curious; he simply never stops asking questions.[13] Fortunately for his sake he is surrounded by a cadre of the very patient: the retired teacher, Mary; the household help, Jonas and Dorothy; the equally curious little brother, Nathan; and a mother and father who are quite obviously well-read in natural philosophy.

Almost all of the chapters in Abbott's science books, those in Table 9.10, are experimental essays for children. By example they teach how to do science as well about scientific matters *per se*. As Abbott explains in the Preface to *Air*, learning about the subject of the essay is secondary to learning about the virtues of an experimental mind set:

---

[13] See Boyle, *Of the Usefulnesse of Natural Philosophy. The First Part. Of its Usefulness in Reference to the Minde of Man.* [38]. Boyle's Pyrophilus is not unlike Abbott's Rollo.

The main design in view in the discussions which are offered to the juvenile world, under the title of THE ROLLO PHILOSOPHY, relates rather to their effect upon the little reader's habits of thinking, reasoning, and observation, than to the additions they may make to his stock of knowledge. The benefit which the author intends that the reader shall derive from them, is an influence on the cast of his intellectual character, which is receiving its permanent form during the years to which these writings are adapted.[2, p. 5]

The experiments in Abbott's science books are described in such Boylean detail that they could readily be performed by a curious reader. For example, in Chapter III of *Air*, Valve Making, Jonas, Rollo, and Nathan repair a bellows.

So Jonas took some leather and cut out a piece of an oblong shape, a little wider than the hole, and about twice as long. Then he laid this down over the hole. It covered it entirely. Then he took some small carpet nails, and nailed one of the ends of the leather down to the board. Then Jonas put his hand down under the board, and ran one of his fingers up through the hole, and pushed the leather up a little way. [2, p. 46]

There is even an experiment in *Fire* about making gunpowder:[14]

"Seventy-five parts of saltpetre," said his father, reading out of his great book, and appearing not to pay any attention to what Rollo said, "seventy-five parts of saltpetre, eleven and a half of sulphur, and thirteen and a half of charcoal. Seventy-five to eleven, that is, about seven to one. Say, six times as much saltpetre as of each of the other two. That will be near enough."[1, p. 91]

Also in accordance with Boyle's guidance for experimental essays—that they must report failures as well as successes—some of Abbott's tales include disappointment at the outcome of an experiment. The following is from an experiment in *Air* in which the flame of a burning wick was supposed to be extinguished when the wick was put into a jar of choke damp:

By this time Jonas thought that the tumbler was filled with the gas, which was rising from the chalk and vinegar. So he rolled up a piece of paper, and set the end on fire, and, when it was well burning, he plunged the end of it into the tumbler. To Rollo's great disappointment and mortification, it continued to burn about as much as ever.

---

[14] ...which may have had to be conducted with a little adult supervision.

[Rollo] looked disappointed and vexed, and appeared to be overwhelmed with chagrin.

Dorothy continued to laugh at them, while Jonas went to the pump and washed out the tumbler. At length she said, —

"But come, Rollo, don't be so disconsolate. You look as if you had swallowed all the choke damp."

"Yes, Rollo," said Jonas, "we must keep good-natured, even if our experiments do fail." [2, p. 137–138]

And if such essays be but as they should be competently stocked with experiments, it is the reader's own fault, if he be not a learner by them.

A Proëmical Essay
Robert Boyle

Chapter 10

Conclusion

In 1817, the editors of the *North-American Review* wrote:

> The library of the Boston Athenaeum contains 11,600 volumes. This collection is rich in many splendid works of natural history....The library of the American Academy is also deposited there, so that there are now in the same building 18,000 volumes. The library of the American Academy contains 1400 volumes, principally works of science, transactions of foreign societies, &c. [80, p. 431]

At the time the library of the Boston Athenæum was among the top five libraries in America by number of volumes.

The promoters of the Anthology Reading-Room were keenly aware of the importance and difficulty of accessing books and periodicals in antebellum America:

> In this country nothing can exceed the inconvenience, arising from the want of large libraries to those persons, who aim at superiour [*sic*] attainments and accurate researches....As much time as is necessary for reading a particular book, is often consumed in attempts to discover or obtain it; and frequently, after every inquiry, the book wanted cannot be procured. [210, pp. 11–12]

The success of the reading room in Joy's Building as well as its successor, the Boston Athenæum, was due in no small part to the formation and maintenance of a collection of resources—books, newspapers, journals, maps, laboratory equipment, and even a cabinet of curiosities—that was consciously and continuously tuned to the interests of the membership. In the document soliciting subscriptions to the Boston Athenæum the manner in which needs of these readers would be served was illustrated as follows:

1. The inquisitive merchant must prize the opportunity of being able to consult a large collection of those works, which relate to commerce.

2. The researches of those who attend to the constitution of society must be greatly facilitated by the assemblage, in one place, of the best and newest treatises upon these subjects of inquiry, of statistical tables and works, and state papers.

3. The historian, and the reader of history, will here be able to perfect their information by a recourse to standard works of general and particular history.

4. Gentlemen of each of the learned professions must derive important assistance from the liberty of consulting both those fugitive and periodical publications and also those large, valuable, and expensive works, which it may be inconvenient to most individuals to purchase. [210, pp. 10–11]

The names of the readers in experimental philosophy at the Boston Athenæum which appear in the preceding chapters also appear in the lists of the founders, trustees, officers, and committee members of Boston's learned societies of the day, from the Mechanics Institute and the Massachusetts Society for Promoting Agriculture to the Boston Society of Natural History and the American Academy of Arts and Sciences. These individuals published articles in the professional and scholarly literature from the *New England Farmer* and the *Franklin Journal* to the *American Journal of Science* and the *Transactions of the American Philosophical Society*. They wrote books of local interest and international stature. They gave public talks, lectures and demonstrations. They taught in Boston's schools and colleges. They built railroads and mill complexes. They were granted patents for their inventions. They conducted surveys for the government. And they even exhibited at the Brighton Cattle Show. Judging from the Athenæum's books borrowed registers there was no scientific or technical topic in which they weren't interested and involved.

In the preceding chapters I have endeavored to gain some insight into the manner in which these individuals practiced science and technology by analyzing their preferences in scientific and technical literature. The first conclusion to which I have come is that the interest of these readers in science and technology was driven in no small part by what they were doing. This is not remarkable but it frames a second, perhaps less obvious, conclusion: they engaged science and technology by example. That is, they preferred a style of scientific and technical exposition that was organized around reports of the author's own experiences.

Further to this second conclusion, as it was often their intent to use what they read in a project in which they were currently involved, Athenæum readers preferred texts that included details of how to adapt the teachings of the text to the particulars of that project. Recall, for example, how Tredgold solved a heating problem using Newcastle coal and cast iron pipes. At each step along the way Tredgold gave the reader all the information needed to tailor his object lesson to an alternative situation; a different kind of fuel, a different type of pipe, a different use of the room, and so forth.

A third conclusion to which I have come is that the readers were interested as much in how and why an author's solution worked as they were in the fact that it did work. They understood that a successful adaptation of the author's solution to a project at hand would not be a rote, tab-A-into-slot-B exercise. As a consequence, it was equally important that the author explain how a solution worked, not in abstract terms but terms of a first-hand, at-scale experience. This understanding, the readers knew, would contribute to the success of their project and could furthermore serve as the basis for elaboration and innovation.

I have called attention to a similarity between the form of writing preferred by readers in experimental philosophy at the Boston Athenæum and Robert Boyle's experimental essay. The scientific and technical texts favored by Athenæum readers centered on specific cases, included all details required to replicate the author's experience, reported successes and failures, and did not speculate on underlying causes. What the essays in the Athenæum texts added to Boyle's requisites was the casting of the discourse in the vocabulary of an application context and provision of explicit means to adapt the outcomes of an experiment to alternative experimental conditions. Thus, for example, in Tredgold's discourse on heating, the solution was expressed in units of pounds of coal. Unlike Boyle's experimental essay, the practitioner's experimental essay was outward looking, building connections between a problem's solution and the context within which it would have to work.[1]

Fourth and finally, I have concluded that the Boston Athenæum's readers were fully abreast of the the scholarly scientific literature of the day. Recall, for example, Uriah A. Boyden reading Euler's papers on the effect of turbulence on water wheel efficiency while improving the hydraulic turbine. At the same time, however, readers paid close attention to the experiences of other practitioners in putting scholarly findings to work. Call it Yankee skepticism or Yankee pragmatism, if you like. They knew that scientific theories could misguide as well as guide. They were aware that effects that were ignored by savants in the laboratory could be controlling in the field.

Conclusions about readers in experimental philosophy at the Boston Athenæum based on analyzing the books and journals they borrowed cannot, of course, be extended beyond that particular population without further work. This study has tried

---

[1] For more about the orthogonal intents of scientists and engineers see Edwin Layton's "Mirror-Image Twins: The Communities of Science and Technology in 19th-Century America" [160]. On page 567 therein Layton notes "It is, of course, very difficult to discover which works were used read by specific inventors; it is even harder to establish a correlation between particular inventions and prior published information." As the preceding strives to demonstrate, difficult but not impossible.

to demonstrate that library charging records can be used to explore how individuals in a well-defined community—in the case at hand, Boston's antebellum practitioners and technologists—acquired, used, and expanded the store of scientific knowledge.

There are many insights left in the Boston Athenæum's books borrowed registers both with respect to readers in science and technology as well with respect to other reader communities. There are digitizations of charging records of other libraries that deserve exploration for what they tell us about early American scientists, engineers and inventors.

The X patents are by no means exhausted as a primary source, particularly when it is recalled that these documents were often written by the inventor in person and, at least from 1793 to 1836, were granted on application and thus reflect an unfiltered, ground-level view of technology of the day.

Digitizations of letters and other personal papers of antebellum technologists such as Loammi Baldwin are coming online and merit examination by scholars of all stripes. The Smithsonian Institution is one of many organizations supporting crowdsourced transcription projects and the output of these projects is certainly worthy of attention.

This book is itself a collection of short experimental essays in bibliography. It is deliberately so. Each vignette is the report of an experiment in connecting what somebody was reading to what they were doing. When the experiment was successful the outcome was insight into both books and their readers.

Barrett Wendell's readers in the humanities were building the literary, artistic, and social infrastructure of the new nation. The readers in experimental philosophy were building the nation's physical, scientific, and technical infrastructure. They were all, it seems to me, reading in no small part for their new nation's practical purposes.

Appendix A

Catalogs, Volumes, Readers, and Charges

If one is to consider books borrowed, it is necessary to inquire what was on offer on the shelves and available for borrowing. Seven catalogs of the Athenæum's library collection were published in the period spanned by the first four volumes of the Athenæum's books borrowed registers These seven catalogs are listed in Table A.1.

A.1   The Boston Athenæum's Printed Catalogs, 1807–1849

Year	Title
1810	Catalogue of the books in the Boston Athenaeum
1810	Catalogue of tracts, scientific and alphabetic index
1827	Catalogue of books in the Boston Athenaeum: to which are added the by-laws of the institution, and a list of its proprietors and subscribers
1829	Catalogue of books added to the Boston Athenaeum since the publication of the catalogue in January 1827
1831	Catalogue of tracts, scientific and alphabetical index
1834	Catalogue of books added to the Boston Athenaeum in 1830–1833
1840	Catalogue of books added to the Boston Athenaeum, since the publication of the catalogue in January, 1827
1849	Shelf Lists, 1849

Table A.1: Printed Catalogs of the Boston Athenæum, 1810–1849

A.2   Classes of the 1810 Catalog

The first mention of the Athenæum's collection of books appears in Articles of Incorporation signed by Governor Caleb Strong on February 13, 1807[65, pp. 20–21]:

> Whereas, the persons hereinafter named, together with sundry other persons, and their associates, have, at very considerable expense collected a library, consisting of rare and valuable books; and, whereas the laudable object of their association is to form, as far as their funds shall from time to time admit, a still more valuable and extensive collection of such rare and valuable works in ancient and modern languages, as are not usually to be met with in our country, but which are deemed indispensible [sic] to those who would perfect themselves in the sciences.

The first mention of a catalog appears in a set of by-laws for the operation of the Athenæum adopted at the April 7, 1808, meeting of the Trustees. Article 17 states

A complete catalog of all the books and pamphlets belonging to the Institution shall be constantly in each of the apartments for the inspection of visitants.

Article 23 says that the creation of the catalog is the responsibility of the Librarian, William Smith Shaw at the time, but Article 24 says that the Librarian can delegate this responsibility. At the meeting on January 1, 1810, the first catalog of the Boston Athenæum was brought into being:

> The Committee for the quarter ending this day beg leave respectfully to report to the Trusteed that they have requested the Rev. Joseph McKean to make a Catalogue of the Books &c. belonging to the Athenæum to which business he has attended and the first part of the Catalog is completed and arrangements are made by the Committee to publish it without delay.

The titles in the 268-page 1810 catalog were divided into the fifteen classes listed in Table A.2. There were seventy-one titles in the *Class IV. Mathematics and Natural Philosophy*.

Class	Subjects	Pages	Entries
I.	Theology, Metaphysics, Ethics, and Ecclesiastical History	1–9	131
II.	Law, Politics, Economics, and Commerce	10–28	299
III.	Medicine and Physiology	29–30	25
IV.	Mathematics and Natural Philosophy	31–35	72
V.	Chemistry and Mineralogy	36–37	22
VI.	Natural History and Botany	38–41	49
VII.	Arts and Sciences; Encyclopedias; Memoirs and Transactions of Learned Societies; Journals of Philosophy and the Arts; Treatises on Agriculture, War, and the Mechanic Arts	42–51	142
VIII.	Fine Arts, Books of Engravings, &c.	52–58	101
IX.	Greek and Latin Authors, with Translations and Illustrations; Lexicons and Grammars of the Antient [sic] Languages; Bibliography and Mythology	59–70	164
X.	Dictionaries and Grammars of Modern Languages; Belles Lettres, Criticism, Rhetoric, Logic, and Books on Education	71–76	72
XI.	History, Biography, Topography, Voyages and Travels; with Maps and Charts.		
	Chap. I. Foreign	77–102	392
	Chap. II. American	103–112	137
XII.	Poetry and Dramatic Works	113–119	93
XIII.	Miscellanies	120–129	154
XIV.	Periodical Works, Journals, Magazines, Newspapers, Reviews, &c.		
	Chap. I. Foreign	130–136	99
	Chap. II. American	137–142	78
XV.	Tracts	143–266	1,985
			4,015

Table A.2: Classes of the 1810 Catalog

A.3 Plots of Volumes, Readers, and Charges

Figure A.1 plots the approximate number of volumes in the Boston Athenæum collection from 1827 to 1850. These counts are taken from the annual librarian's report. Figure A.2 plots the number of readers and Figure A.3 plots the number of charges over the same time period. These counts are taken from the books borrowed registers.

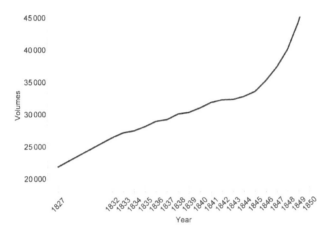

Figure A.1: Plot of Volumes in the Athenæum's Collection, 1827–1850

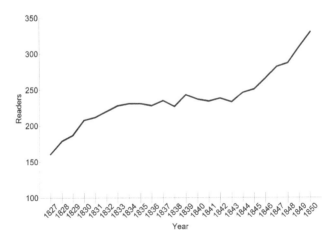

Figure A.2: Plot of the Number of Readers, 1827–1850

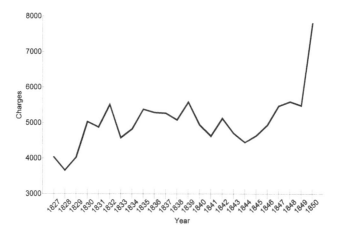

Figure A.3: Plot of the Number of Charges, 1827–1850

A.4   The Lost Catalog of the Massachusetts Scientific Association

The Massachusetts Scientific Association was perhaps the shortest lived of all ante-bellum scientific associations. The record of its proceedings is only eighteen pages long. [208] In a previous report of the Bowditch committee printed on May 25, 1826, regarding their charge to examine deficiencies in the library we find the following:

> The Scientific Association, formed a few months since, for the purpose of procuring a collection of scientific books, have obtained a subscription of 3,715 dollars, of which above 3,000 dollars have already been collected. An agreement for a union with this association, having been made by the committee, was approved by the Proprietors of the Athenaeum, at their meeting May 25th, 1826, and the whole amount subscribed will be appropriated for the purchase of the scientific books contained in a catalogue prepared by the Trustees of the Scientific Association. This catalogue does not contain any books now in the Athenæum, and it will make a very important addition to their already respectable collection of scientific books. This department of the Athenaeum will also be rendered much more complete by the sum subscribed in February, 1826, for completing the transactions of the Royal Societies and Academies of Sciences in London, Edinburgh, Dublin, Paris, Petersburg, Berlin, Turin, Göttingen, Stockholm, Copenhagen, Madrid, and Lisbon, making, in the whole, one of the most complete scientific libraries in the United States.

It would seem that the Massachusetts Scientific Association or at least its library was destined to be assimilated by the Athenæum even as it was being formed. What was deposited with the Athenæum was not books but money—$3,715.00 to be exact— that had been collected to buy scientific books, periodicals, and instruments. Judging from the list of subscribers, the Massachusetts Scientific Library Association, much like Boston's Museum of Fine Arts a few years later [134], was conceived in the quiet of the Athenæum's reading room.

The list of periodicals together with the vision of becoming "one of the most complete scientific libraries in the United States" testifies to the wide-ranging awareness and interest of Athenæum readers in science of the day as well as their understanding of the connection between science and technology.

Appendix B

Charge Counts for All Readers in Volumes I–IV

Reader	Charges	First Charge	Last Charge
Abbot, Lawrence	1	2/1/1845	2/1/1845
Abbot, Samuel L.	168	8/1/1843	1/25/1850
Abbott, Jacob	97	1/1/1839	10/11/1850
Adams, Abigail Brooks	12	1/13/1851	2/10/1851
Adams, Benjamin	103	2/3/1827	1/20/1831
Adams, Charles Francis	750	1/4/1830	2/3/1851
Adams, Charles Frederick	47	2/8/1845	2/21/1846
Adams, Francis M.	50	3/19/1849	6/18/1850
Adams, Hannah	22	4/3/1827	5/3/1828
Adams, L.B.	112	2/2/1827	12/21/1829
Adams, Zabdiel B.	245	1/7/1830	1/22/1851
Alcott, Bronson A.	18	1/13/1849	2/11/1851
Alexander, Andrew	91	7/25/1848	10/18/1851
Alger, Cyrus	98	1/17/1829	12/20/1844
Alger, Francis	445	6/30/1843	2/11/1851
Allen, Robert B.	9	4/12/1833	12/20/1833
Amory, Charles	102	2/5/1847	2/10/1851

Reader	Charges	First Charge	Last Charge
Amory, Thomas Coffin, Jr.	377	6/2/1846	1/27/1851
Amory, William	10	3/24/1849	11/25/1850
Anderson, Rufus	90	3/20/1834	9/23/1842
Andrews, Caleb	207	1/4/1832	12/20/1842
Andrews, James	203	8/28/1827	8/12/1836
Andrews, William Turell	408	1/7/1831	2/11/1851
Appleton, Edward	16	2/10/1827	11/13/1827
Appleton, Francis Henry	5	2/6/1845	2/6/1846
Appleton, Nathan	205	2/16/1827	12/12/1850
Appleton, Samuel	102	2/15/1827	11/19/1850
Appleton, William	161	3/22/1827	1/27/1851
Arkin, Silas	2	6/7/1839	6/7/1839
Armstrong, Samuel Turell	40	4/1/1830	1/12/1835
Arvine, K.	7	5/30/1850	11/30/1850
Atkinson, Edward	289	8/1/1846	2/10/1851
Atkinson, William Parsons	407	1/1/1844	2/10/1851
Atwood, Charles	392	11/19/1838	2/10/1851
Austin, Henry David	87	1/27/1849	2/4/1851
Austin, James Trecothick	698	2/7/1827	12/27/1843
Austin, Samuel	6	9/28/1849	11/16/1850
Austin, Samuel, Jr.	120	3/13/1841	9/8/1848
Aylwin, William Cushing	56	7/4/1845	8/22/1848
Bacon, Daniel Carpenter	1	7/20/1846	7/20/1846
Baker, A.B.	49	11/23/1846	11/11/1847
Ball, Stephen	19	4/27/1850	12/19/1850
Bancroft, George	236	8/14/1833	1/29/1845
Bangs, George Pemberton	6	8/30/1850	1/20/1851
Barnard, Charles	1009	2/6/1827	2/3/1851
Barnes, Isaac Orr	6	4/12/1850	12/3/1850
Barrell, Samuel B.	94	1/10/1839	8/31/1842
Barrett, Samuel	111	8/14/1829	4/15/1847
Bartlett, Sidney	78	3/12/1845	1/8/1851

Reader	Charges	First Charge	Last Charge
Bartlett, Thomas	198	1/30/1827	6/7/1844
Bartol, Cyrus Augustus	472	2/9/1837	2/4/1851
Bass, Seth	1	10/24/1849	10/24/1849
Bassett, Francis	22	4/3/1841	7/25/1850
Batch, Joseph	20	7/30/1849	1/28/1851
Bates, George	778	2/22/1827	2/7/1851
Beck, Charles	24	1/1/1845	3/30/1850
Becker, Mrs.	35	1/9/1850	12/24/1850
Beecher, Edward	11	4/1/1846	6/12/1846
Belknap, Andrew Eliot	54	8/17/1844	11/18/1850
Belknap, John	1078	2/9/1827	1/27/1851
Bell, Joseph	332	3/16/1843	1/8/1851
Bemis, Samuel A.	7	3/25/1829	4/20/1830
Bethune, John McLean	819	1/3/1842	2/10/1851
Bigelow, Erastus Brigham	5	10/18/1850	12/21/1850
Bigelow, Henry Jacob	3	10/24/1850	11/20/1850
Bigelow, Jacob	813	2/6/1827	2/3/1851
Binney, Amos	44	7/24/1833	4/19/1849
Binney, Amos, Mrs.	25	8/2/1849	12/24/1850
Bird, John A.	27	2/24/1849	3/25/1850
Blagden, George W.	115	6/23/1834	2/5/1851
Blake, Edward	114	1/15/1836	4/14/1842
Blake, Joshua	6	3/3/1840	10/3/1840
Blake, Samuel Parkman	46	4/23/1834	1/1/1849
Blake, William	120	1/15/1845	12/21/1850
Blanchard, Edward	6	3/14/1849	9/6/1850
Blanchard, John Adams	223	3/23/1846	2/7/1851
Bliss, William Davis	27	1/12/1850	1/23/1851
Bogen, Frederic W.	19	6/14/1850	1/13/1851
Bokum, Hermann	7	4/16/1836	4/12/1837
Bond, George	92	2/14/1829	3/19/1835
Booth, C.H.	8	10/2/1845	11/18/1845

Reader	Charges	First Charge	Last Charge
Boott, Francis	86	5/4/1838	12/27/1842
Boott, John W.	199	1/10/1829	1/18/1844
Boott, Kirk	82	3/12/1829	3/24/1837
Borland, John Nelson	18	12/31/1850	12/27/1851
Bowditch, Henry Ingersoll	219	4/17/1839	12/3/1850
Bowditch, Jonathan Ingersoll	288	7/1/1836	2/6/1851
Bowditch, Nathaniel	254	2/27/1827	3/13/1838
Bowditch, Nathaniel Ingersoll	375	2/11/1834	2/1/1851
Bowdoin, James	10	8/25/1827	4/20/1828
Bowen, Francis	160	6/12/1844	2/7/1851
Bowen, Francis, Jr.	31	1/13/1843	6/10/1844
Boyden, Uriah Atherton	73	3/19/1841	1/1/1851
Bradbury, Charles	403	2/22/1827	12/20/1842
Bradford, Thomas G.	140	1/14/1837	6/10/1839
Bradlee, Frederick Wainwright	9	12/4/1849	1/31/1851
Bradlee, James Bowdoin	20	8/2/1848	8/14/1850
Bradlee, Josiah	428	1/10/1829	4/21/1849
Bradlee, Nathaniel Jeremiah	7	7/5/1850	11/30/1850
Bradlee, William C.	4	5/18/1850	10/23/1850
Brewer, Gardner	76	8/4/1849	4/20/1851
Brewer, Thomas Mayo	2	1/1/1849	1/23/1851
Bridge, Samuel James	25	5/26/1832	12/24/1832
Briggs, William A.	17	10/12/1842	7/5/1843
Brigham, Elijah Dana	1	11/8/1850	11/8/1850
Brimmer, George W.	105	3/7/1827	10/23/1835
Brimmer, Martin	105	4/21/1830	9/26/1850
Brooks, Benjamin	118	4/15/1841	7/3/1844
Brooks, Edward	434	8/14/1828	10/25/1847
Brooks, Francis Augustus	34	1/5/1850	2/8/1851
Brooks, Francis Boott	2	10/28/1850	12/4/1850
Brooks, Gorham	672	8/29/1828	12/28/1850
Brooks, Peter Chardon	206	3/26/1834	1/1/1849

Reader	Charges	First Charge	Last Charge
Brooks, William Gray	48	1/21/1842	12/14/1842
Brooks, William H.	124	8/4/1845	1/26/1851
Brown, John	7	7/27/1830	12/28/1830
Bryant, John	90	4/16/1828	3/6/1847
Buckingham, Joseph	1	2/13/1850	2/13/1850
Bullard, William Story	122	2/27/1840	12/28/1850
Bumstead, John	245	2/8/1827	11/17/1840
Burley, Susan	1077	2/23/1837	12/31/1850
Burroughs, George	752	7/12/1834	11/19/1847
Butler, Clement M.	36	10/1/1844	1/2/1847
Cabot, Samuel	182	1/10/1835	2/3/1851
Cabot, Samuel, Jr.	154	2/5/1839	2/8/1851
Cabot, Thomas Handasyd	25	1/21/1831	12/15/1831
Capen, Nahum	201	10/16/1828	3/1/1841
Carrire, Jules J.	3	2/1/1851	2/5/1851
Cary, Ann M.	31	2/4/1850	11/4/1850
Cary, Thomas Greaves	381	4/4/1832	2/10/1851
Casenove, Charles John	93	2/13/1827	8/20/1832
Chace, Caleb	9	7/31/1850	1/11/1851
Chadwick, Ebenezer	14	1/28/1847	1/23/1849
Chandler, Abiel	142	2/10/1827	12/9/1831
Chandler, Gardiner L.	152	12/28/1844	1/20/1851
Chandler, Gardiner L., Jr.	268	1/7/1835	8/12/1837
Channing, Walter	303	3/10/1827	10/3/1839
Channing, William Ellery	261	3/6/1827	2/11/1851
Chapman, George	116	2/28/1831	8/21/1832
Chapman, Jonathan	400	3/24/1827	6/16/1847
Chapman, M.W., Mrs.	356	1/15/1847	2/10/1851
Chapman, O.G.	41	8/14/1849	1/28/1851
Chase, Theodore	144	2/23/1848	2/1/1851
Chase, William M.	51	3/1/1850	12/4/1850
Child, D.L., Mrs.	92	1/14/1832	2/2/1835

Reader	Charges	First Charge	Last Charge
Choate, Rufus	584	5/29/1839	2/12/1851
Clark, John	7	2/6/1827	6/18/1827
Clarke, Benjamin Cutler	18	12/24/1850	2/11/1851
Clarke, James Freeman	358	1/1/1839	7/24/1850
Clinch, Joseph H.	89	1/8/1847	2/28/1851
Coale, W.C.	16	1/7/1846	11/7/1846
Cobb, Frederick Augustus	72	1/10/1830	6/30/1832
Cobb, Richard	311	1/1/1830	8/2/1836
Codman, Henry	210	4/11/1827	11/20/1850
Coffin, George Washington	36	1/19/1847	1/27/1851
Coffin, John Gorham	39	2/3/1827	8/21/1828
Coffin, Joy	296	1/12/1839	7/19/1844
Coffin, Thomas M.	123	7/19/1844	1/30/1851
Colburn, Warren	78	3/28/1827	1/17/1833
Cole, Dr.	31	11/23/1840	11/18/1843
Colman, Henry	26	8/1/1838	10/19/1843
Cooke, Josiah Parsons	230	5/25/1842	1/1/1849
Cooke, Josiah Parsons, Jr.	193	7/18/1844	12/21/1848
Coolidge, John T.	32	1/1/1848	1/21/1851
Coolidge, Joseph, Jr.	182	3/31/1827	2/13/1834
Coolidge, Samuel F.	14	2/17/1827	1/18/1828
Cordis, Thomas	21	2/17/1827	1/9/1850
Cort, Daniel T.	52	3/12/1829	2/15/1831
Cort, Thomas W.	3	10/23/1827	11/15/1827
Cotting, B.E.	41	4/21/1844	12/28/1848
Courties, Ambrose S.	85	1/12/1830	11/21/1832
Cruft, Edward	1169	2/5/1827	2/10/1851
Cunningham, Caleb Loring	12	2/9/1850	1/22/1851
Curtis, Benjamin Robbins	76	2/4/1843	1/21/1851
Curtis, Caleb	122	8/10/1844	2/5/1851
Curtis, Charles Pelham	532	2/6/1828	2/3/1851
Curtis, Nathaniel, Jr.	29	7/3/1841	4/20/1842

Reader	Charges	First Charge	Last Charge
Curtis, Thomas Buckminster	178	2/13/1837	1/31/1851
Cushing, Caleb	178	10/14/1828	10/17/1839
Cushing, Luther Stearns	27	1/1/1849	12/11/1850
Cushing, Thomas Parkman	31	7/1/1846	2/4/1851
Dana, Richard Henry, Jr.	15	1/1/1850	2/10/1851
Dana, Samuel, Sr.	3	8/10/1827	9/10/1827
Danforth, Bowers	1446	2/17/1827	2/7/1851
Darling, Benjamin	18	4/7/1827	12/31/1827
Darrah, Robert K.	10	8/9/1848	1/1/1850
Davis, Edward Gardiner	347	11/6/1828	10/11/1834
Davis, Francis	9	4/3/1846	9/22/1846
Davis, Isaac P.	885	2/1/1827	11/16/1846
Davis, John	330	2/3/1827	9/16/1846
Davis, John Brazer	263	2/19/1827	11/23/1832
Davis, Jonathan	113	2/12/1827	11/4/1828
Davis, Joshua William, Mrs.	47	11/5/1847	7/8/1848
Davis, Thomas	82	1/3/1833	3/25/1840
Dearborn, Henry Alexander Scammell	91	2/17/1827	6/12/1830
Degrand, Peter P.F.	5	1/31/1839	7/27/1840
Derby, Richard C.	7	2/15/1827	4/12/1827
Dexter, Franklin	639	2/7/1827	2/7/1851
Dexter, George Minot	151	11/11/1828	1/27/1851
Dillaway, Charles Knapp	213	1/18/1835	12/16/1841
Dix, J.H.	10	4/15/1843	11/24/1843
Dixon, Thomas	3	8/2/1849	8/30/1849
Dixwell, Epes Sargent	115	3/21/1835	12/30/1841
Dixwell, John James	246	3/17/1827	10/9/1850
Doane, George W.	30	1/29/1829	8/6/1832
Dodge, Pickering	41	10/23/1834	12/13/1849
Doggett, John	3	2/10/1827	2/10/1827
Dorr, Charles Hazen	141	7/9/1844	11/23/1850
Dorr, John	51	1/1/1836	9/25/1837

Reader	Charges	First Charge	Last Charge
Dorr, Samuel	422	2/16/1827	7/21/1844
Downer, Samuel	13	8/4/1849	10/23/1850
Downer, Samuel, Jr.	72	7/31/1841	3/30/1849
Dwight, Edmund	269	2/26/1831	12/7/1850
Dyer, Henry	13	1/21/1843	9/6/1843
Eaton, Asa	344	1/20/1837	12/23/1843
Eddy, Robert Henry	226	1/1/1836	11/30/1848
Edwards, B.B.	208	3/20/1834	12/30/1840
Edwards, Henry	90	1/6/1846	12/14/1846
Eldridge, Edward Henry	81	5/28/1839	12/27/1841
Eldridge, Oliver	70	3/4/1847	1/30/1851
Eliot, Samuel	9	8/3/1850	1/31/1851
Eliot, Samuel Atkins	49	10/16/1827	1/16/1851
Eliot, William Havard	6	1/8/1828	12/10/1828
Ellis, David	727	1/18/1834	12/22/1838
Emerson, George Barrell	621	2/22/1827	1/29/1851
Emerson, Ralph Waldo	341	1/1/1830	1/31/1851
Everett, Alexander H.	253	1/11/1830	8/31/1840
Everett, Edward	120	7/12/1827	5/8/1840
Fairbanks, Charles B.	190	7/3/1847	2/10/1851
Fairbanks, Gerry	231	3/17/1827	12/8/1829
Fairbanks, Stephen	247	1/6/1847	2/7/1851
Fales, Samuel	2	3/30/1848	4/8/1848
Farnham, Luther	108	12/21/1849	2/11/1851
Farnsworth, Amos	74	2/1/1827	7/7/1832
Farrar, George	91	7/1/1846	8/6/1849
Faxon, Nathaniel	341	1/12/1837	12/20/1845
Felt, Joseph Barlow	1130	2/20/1827	2/3/1851
Felton, Cornelius Conway	22	2/5/1838	12/5/1839
Fisher, Francis	75	4/1/1846	12/3/1850
Flagg, Josiah Foster	216	2/2/1827	2/11/1851
Fletcher, Richard	492	1/19/1828	2/13/1851

Reader	Charges	First Charge	Last Charge
Folsom, Charles	1321	3/16/1846	2/14/1851
Forbes, John Murray	27	12/8/1846	11/16/1850
Forbes, Robert Bennet	68	1/4/1833	12/23/1837
Foster, John	59	1/15/1830	3/19/1834
Foster, Thomas	112	2/3/1827	1/7/1830
Foster, William	215	2/5/1827	8/22/1833
Fox, Thomas B.	19	3/28/1848	5/4/1849
Francis, Ebenezer	150	2/5/1827	2/11/1851
Freeman, James	149	2/26/1828	3/21/1836
Freeman, James, Jr.	163	2/5/1828	1/10/1835
French, Jonathan	7	1/25/1845	12/6/1845
Frothingham, Nathaniel Langdon	1356	1/2/1829	2/5/1851
Gallup, George G.	5	7/2/1847	7/15/1848
Gardiner, Chandler L., Jr.	171	8/25/1832	1/14/1835
Gardiner, William Howard	71	6/30/1832	11/14/1850
Gardner, Brewer	88	1/11/1845	4/2/1849
Gardner, George	3	7/25/1846	5/2/1850
Gardner, Henry	68	3/20/1841	2/26/1843
Gardner, John Lowell	1	3/11/1848	1/1/1849
Gardner, Samuel Jackson	17	4/10/1827	12/26/1833
Gardner, Samuel Pickering	5	4/2/1827	12/27/1827
Gay, Martin	187	3/30/1829	9/8/1849
Gibbs, William	55	4/3/1839	12/10/1840
Gilchrist, Daniel Swan	84	8/12/1847	2/10/1851
Gilchrist, Edward	68	1/9/1850	1/30/1851
Goodrich, Samuel Griswold	328	7/22/1830	1/9/1840
Gore, John Christopher	48	9/11/1830	2/17/1841
Gorham, Benjamin	35	11/30/1844	3/7/1846
Gorham, John	181	3/29/1827	8/27/1838
Gould, Augustus Addison	238	6/19/1832	6/14/1850
Gragg, Washington Parker	105	3/28/1827	12/24/1829
Gray, Francis Calley	342	3/13/1828	2/3/1851

Reader	Charges	First Charge	Last Charge
Gray, Horace	316	2/12/1827	9/18/1847
Gray, Horace, Jr.	89	2/22/1848	2/11/1851
Gray, John Chipman	63	2/12/1827	12/13/1850
Gray, Samuel Calley	311	1/29/1827	1/1/1849
Gray, Thomas	5	10/1/1850	1/13/1851
Gray, William Rufus	113	4/4/1827	8/19/1828
Green, H.B.C.	1	5/10/1843	5/30/1843
Greene, Benjamin Daniel	142	2/10/1827	6/17/1850
Greene, Gardiner	112	3/8/1828	2/28/1830
Greene, Gardiner, Mrs.	131	7/19/1827	7/1/1833
Greene, Nathaniel	78	9/22/1832	3/5/1850
Greenleaf, Ezekiel Price	275	10/21/1844	12/6/1850
Greenleaf, Simon	8	9/9/1846	4/9/1847
Greenough, Alfred	157	1/4/1839	2/8/1851
Greenough, William Whitwell	59	1/1/1835	2/10/1851
Greenwood, Francis William Pitt	15	1/1/1827	5/30/1846
Grew, Henry	23	1/9/1833	10/13/1834
Grigg, William	380	1/21/1828	12/27/1834
Guardenier, John	403	7/24/1833	9/12/1849
Guild, Benjamin	377	1/1/1827	1/28/1851
Hague, William	144	1/7/1833	5/8/1848
Hale, Enoch	718	2/9/1827	11/6/1848
Hale, Nathan	258	7/6/1827	2/11/1851
Hall, James	210	4/12/1830	1/10/1835
Hall, Joseph	698	2/17/1827	3/10/1848
Hall, Nathaniel	31	1/29/1846	4/15/1847
Hall, Thomas Bartlett	115	1/16/1846	12/11/1849
Hallet, Henry S.	1	11/30/1850	11/30/1850
Hammond, Daniel	1659	1/7/1829	10/19/1850
Hancock, John	1124	2/3/1827	2/8/1851
Hancock, John, Jr.	95	4/12/1839	12/9/1842
Hancock, John, Sr.	192	6/12/1828	2/7/1835

Reader	Charges	First Charge	Last Charge
Harris, Henry	19	2/1/1827	9/29/1827
Harris, Thaddeus Mason	225	3/4/1829	3/3/1841
Harris, Thaddeus William	57	12/13/1838	1/21/1851
Hastings, Daniel	115	2/7/1827	4/23/1836
Havey, Charles F.	26	2/28/1845	3/30/1849
Hawes, Prince	33	11/28/1847	12/22/1848
Hayes, Francis Brown	34	1/6/1847	2/8/1851
Hayward, George	312	2/7/1827	10/22/1850
Hayward, James	9	1/1/1842	12/5/1842
Hayward, Joshua H.	52	3/10/1827	11/15/1831
Healy, Mark	112	11/13/1848	2/5/1851
Heard, John, Jr.	387	1/1/1827	12/21/1839
Heaven, Franklin	7	2/19/1846	4/25/1846
Henderson, A.A.	15	4/6/1849	9/29/1849
Henshaw, David	231	1/1/1827	12/7/1841
Henshaw, Samuel	1	3/7/1827	3/7/1827
Hesmer, Z.	43	6/8/1836	5/3/1837
Hicks, James Henry	54	1/2/1839	11/23/1839
Higginson, Francis	104	1/17/1829	12/17/1830
Higginson, Henry	470	1/9/1830	12/30/1837
Higginson, James Perkins	252	1/20/1843	1/31/1851
Higginson, Waldo	13	7/29/1844	1/6/1845
Hillard, George Stillman	191	6/19/1842	1/6/1851
Himes, Joshua Vaughn	14	11/3/1842	1/12/1843
Hitchcock, Robert S.	18	5/3/1847	9/10/1851
Hobart, Aaron	29	7/1/1842	12/11/1845
Hobbes, Prentiss	3	1/8/1828	1/8/1828
Hoch, H.T.	2	6/20/1844	8/17/1844
Hoit, A.G.	172	1/29/1841	1/29/1851
Holbrook, Henry M.	34	1/26/1850	11/27/1850
Holbrook, Henry Ware	12	1/5/1848	10/20/1848
Holmes, Abiel	22	1/1/1827	11/4/1837

Reader	Charges	First Charge	Last Charge
Holmes, Oliver Wendell	174	10/20/1842	12/16/1850
Homans, I. Smith	25	1/23/1850	1/21/1851
Homer, Charles	1	9/12/1848	1/1/1850
Homer, Joseph Warren	81	4/5/1846	11/14/1850
Hooker, Anson, Jr.	1	2/10/1827	2/10/1827
Hooper, Henry N.	175	2/17/1827	6/22/1833
Hooper, Nathaniel	43	11/30/1849	2/10/1851
Hooper, Robert William	247	6/15/1844	1/25/1851
Hooper, Samuel	355	2/7/1833	1/29/1851
Hovey, Charles Fox	58	8/1/1849	2/8/1851
Howard, John Clarke	41	3/8/1827	10/4/1832
Howe, Estes	10	1/4/1845	7/10/1845
Howe, George	7	1/27/1840	1/8/1851
Howe, Jabez C.	4	8/9/1847	12/9/1850
Howe, Joseph N.	5	4/8/1850	8/7/1850
Howe, Joseph N., Jr.	3	4/11/1845	5/3/1845
Howe, Samuel Gridley	378	10/16/1834	6/29/1850
Howes, William Burley	22	1/30/1851	2/11/1851
Hubbard, J.P.	10	2/26/1841	3/18/1842
Hunt, E.B.	35	12/6/1849	2/8/1851
Hurd, William	290	1/1/1847	2/13/1851
Inches, Henderson	929	1/30/1827	1/8/1851
Ingalls, William	95	7/20/1829	4/12/1850
Ingersoll, James	42	1/4/1845	5/4/1846
Jackson, Charles	297	7/3/1827	2/13/1851
Jackson, James	66	3/10/1827	1/1/1851
Jackson, John Barnard Swett	8	2/4/1843	7/18/1843
Jackson, Lydia	156	5/16/1847	1/29/1851
Jackson, Patrick Tracy	338	6/23/1832	5/1/1847
Jaques, Henry L.	11	9/24/1851	12/29/1851
Jenks, John H.	23	7/14/1832	1/15/1833
Jenks, William	745	2/26/1830	2/8/1851

Reader	Charges	First Charge	Last Charge
Jones, Anna P.	157	3/21/1835	1/23/1851
Jones, J., Mrs.	197	1/15/1842	7/13/1844
Jones, John C.	237	2/15/1827	1/1/1851
Jones, Mrs.	186	6/3/1844	12/3/1846
Joy, Benjamin	21	3/29/1832	2/4/1833
Joy, Joseph B.	14	3/29/1834	10/17/1834
Kendall, Abel, Jr.	53	8/6/1850	12/18/1851
Kent, Moody	102	1/2/1838	12/20/1842
Kettell, Samuel	452	3/23/1842	12/26/1850
King, T.S.	3	9/12/1850	12/27/1850
Kirk, John F.	132	7/9/1846	12/18/1847
Kirkland, John Thornton	42	1/30/1839	5/4/1839
Knowles, Seth	41	1/11/1830	1/8/1831
Kraitser, Charles	5	10/7/1846	12/6/1846
Lamb, Thomas	78	2/17/1827	11/24/1849
Lamson, Alvan	77	6/1/1843	1/28/1851
Lamson, John	27	9/24/1849	11/22/1850
Lawrence, Abbot	129	4/13/1827	12/23/1850
Lawrence, Amos Adams	69	12/23/1840	9/21/1848
Lawrence, William Richards	432	1/24/1832	2/3/1851
Layman, George W.	23	1/7/1837	11/11/1840
Lee, Henry	287	1/1/1829	1/10/1851
Lee, Henry, Jr.	116	12/13/1830	9/20/1850
Lee, Thomas, Jr.	112	2/10/1827	12/6/1830
Lieber, Francis	33	9/20/1828	8/9/1832
Lienon, Henry	8	2/21/1827	6/15/1827
Lincoln, Benjamin	46	7/29/1828	2/8/1851
Lippett, George W.	23	1/12/1850	10/26/1850
Littell, E.	23	1/27/1848	7/28/1848
Little, Charles Coffin	5	8/30/1847	4/20/1848
Livermore, George	270	1/1/1835	1/23/1851
Lloyd, James	70	2/21/1845	12/27/1848

Reader	Charges	First Charge	Last Charge
Lodge, Giles Henry	129	1/1/1836	2/8/1851
Lodge, James	5	5/29/1848	1/6/1849
Lodge, John E.	3	11/3/1839	11/3/1839
Lord, John	90	3/18/1847	9/24/1850
Loring, Benjamin	86	2/4/1833	2/28/1842
Loring, Caleb	192	4/7/1827	12/8/1834
Loring, Charles Greely	20	1/24/1828	2/1/1851
Loring, Edward Greely	203	1/8/1832	8/27/1845
Loring, Elijah	1122	2/19/1834	4/16/1849
Loring, Elijah, Mrs.	13	4/22/1850	11/23/1850
Loring, G.E.	7	12/19/1834	1/2/1835
Loring, Josiah	145	2/10/1827	3/4/1838
Lou, Thomas, Jr.	33	4/20/1839	9/16/1843
Lovering, Joseph Swain	65	11/12/1847	1/23/1851
Lowell, Edward Jackson	99	7/15/1828	9/6/1830
Lowell, Francis Cabot	80	2/16/1827	1/1/1850
Lowell, John	374	2/8/1827	12/8/1840
Lowell, John Amory	185	2/2/1827	1/30/1851
Lowell, John, Jr.	103	2/3/1827	10/13/1833
Lowell, John, Sr.	299	1/7/1830	12/30/1834
Lowell, Rebecca Amory	84	2/18/1841	1/1/1850
Lunt, William P.	476	8/16/1836	1/30/1851
Lyman, George Williams	94	3/29/1831	12/12/1834
Lyman, Theodore	124	1/7/1827	3/11/1848
Lyman, William	622	2/15/1827	2/5/1845
Malcom, Howard	68	1/23/1829	1/7/1831
Martin, Enoch	608	4/29/1837	2/1/1851
Mason and Larthrop	30	1/11/1847	12/7/1847
Mason, Charles	42	9/19/1849	12/2/1850
Mason, Jonathan	2	1/1/1827	1/1/1849
Mason, Robert Means	91	1/14/1845	2/1/1851
Mason, William Powell	26	8/7/1827	2/7/1851

Reader	Charges	First Charge	Last Charge
Mass. Agricultural Soc.	1	11/29/1850	11/29/1850
Mayo, John M.	6	10/4/1850	2/12/1851
McKean, Joseph W.	89	7/28/1828	4/21/1837
Merriam, Charles	45	10/19/1850	2/10/1851
Merrill, Amos B.	32	12/23/1847	1/26/1849
Miles, Solomon Pierson	161	7/7/1827	4/16/1842
Mills, James Kellogg	36	1/1/1848	11/30/1850
Minot, William	29	2/8/1849	11/30/1850
Mixter, Charles	31	11/17/1850	2/1/1851
Moody, David P.	6	12/27/1827	2/12/1833
Moody, Paul	16	5/26/1827	11/6/1828
Motley, Thomas	307	2/13/1827	2/3/1851
Mott, J.M.	982	1/17/1835	12/10/1842
Motte, Mellish J., Jr.	97	1/4/1832	1/14/1835
Neale, R.H.	16	2/2/1839	12/9/1839
Nichols, Benjamin Ropes	610	2/16/1829	12/8/1850
Norton, Charles Eliot	11	1/1/1849	12/1/1849
O'Brien, Carolan	134	10/15/1849	2/5/1851
Odin, John	216	1/19/1828	12/10/1831
Oliver, Eldridge	118	7/1/1844	1/9/1847
Osgood, David	1540	2/13/1827	1/30/1851
Otis, George W., Jr.	437	8/15/1828	12/11/1841
Otis, Harrison Gray	88	2/20/1827	11/26/1845
Page, Henry Augustus	264	8/17/1844	2/12/1851
Paige, T.H.	2	3/3/1827	3/3/1827
Palfrey, J.G.	31	10/28/1841	8/16/1847
Parker, Charles H.	264	8/26/1844	2/8/1851
Parker, Daniel Pinckney	186	5/26/1836	12/21/1850
Parker, Francis Edward	18	3/16/1848	1/28/1851
Parker, George S.	79	8/5/1844	8/3/1846
Parker, James	3	11/16/1849	11/16/1849
Parker, John	2	7/21/1842	7/21/1842

Reader	Charges	First Charge	Last Charge
Parker, John, Jr.	24	1/1/1827	12/15/1837
Parker, Peter	51	1/10/1840	11/9/1841
Parker, Samuel P.	49	7/18/1838	3/28/1840
Parker, Theodore	170	4/8/1846	1/17/1851
Parker, William A.	10	9/10/1850	12/30/1850
Parkman, Charles	188	12/7/1842	3/8/1848
Parkman, Daniel	9	1/10/1828	6/17/1828
Parkman, Francis	1298	2/6/1827	2/7/1851
Parkman, Francis, Jr.	35	4/30/1850	2/12/1851
Parkman, George	1297	2/5/1827	6/22/1844
Parkman, George F., Jr.	503	7/13/1844	2/10/1851
Parkman, John	28	1/1/1827	2/3/1851
Parkman, Samuel	109	5/3/1845	1/2/1851
Parks, W.S.	16	8/23/1849	5/25/1850
Parris, Alexander	638	6/30/1827	3/11/1843
Parrott, William Pearce	84	3/24/1849	2/8/1851
Parsons, Theophilus	453	1/12/1829	2/13/1851
Parsons, Thomas William, Jr.	484	10/3/1837	2/11/1851
Parsons, William	492	3/10/1830	12/28/1850
Payne, Edward William	115	5/31/1827	9/20/1834
Payson, Arthur Lithgow	2	1/13/1848	1/13/1848
Peabody, Ephraim	280	3/15/1841	2/10/1851
Peabody, Miss	23	11/20/1832	12/14/1833
Pelham, Charles	64	7/1/1839	3/4/1843
Perkins, James, Mrs.	203	12/13/1834	11/23/1840
Perkins, Stephen Higginson	568	2/26/1827	12/7/1850
Perkins, Thomas Handasyd	128	3/26/1827	11/6/1850
Perkins, Thomas Handasyd, Jr.	2	1/26/1850	1/29/1851
Perkins, William Powell	47	3/1/1843	11/3/1843
Phelps, Austin	11	4/27/1844	11/14/1844
Phillips, Edward	37	2/28/1827	4/26/1828
Phillips, Jonathan	575	2/17/1827	1/7/1851

Reader	Charges	First Charge	Last Charge
Phillips, Stephen Henry	10	8/31/1847	2/7/1849
Phipps, Samuel	16	1/18/1847	3/16/1850
Pickering, Dodge	132	4/1/1845	4/4/1849
Pickering, Octavius	527	1/1/1830	1/29/1851
Pirscher, Mr.	3	4/12/1835	5/18/1835
Poor, Benjamin	53	3/2/1827	12/18/1827
Pope, Thomas Butler	33	1/30/1840	12/30/1840
Porter, Jonathan	386	1/1/1828	2/10/1851
Pratt, George	219	1/13/1845	1/13/1851
Pratt, William	204	2/11/1827	3/21/1837
Prescott, William Hickling	565	2/3/1827	2/8/1851
Putnam, Benjamin W.	8	12/27/1850	2/1/1851
Putnam, C.G.	23	7/13/1844	12/16/1847
Putnam, Jesse P.	278	2/2/1827	8/10/1839
Putnam, Samuel R.	84	8/21/1835	1/1/1851
Quincy, Edmund	293	2/21/1842	2/7/1851
Quincy, Josiah	537	2/2/1827	2/7/1851
Quincy, Josiah, Jr.	826	1/2/1828	1/2/1851
Rand, Edward Sprague	36	1/30/1839	12/7/1842
Randall, Elizabeth	115	10/30/1844	2/8/1851
Randall, John	1216	3/24/1827	7/2/1844
Read, William	228	2/3/1849	2/8/1851
Reed, Benjamin Tyler	17	6/4/1850	1/22/1851
Reed, Sampson	8	1/22/1847	12/29/1847
Reid, William	40	10/2/1849	11/6/1849
Reilly, Devin I.	3	6/18/1851	6/18/1851
Revere, Joseph Warren	51	3/25/1846	1/28/1851
Reynolds, Edward, Jr.	207	4/2/1830	12/6/1843
Reynolds, Grindall	11	12/17/1850	2/3/1851
Rice, Henry	131	1/7/1836	12/26/1840
Rice, Henry Gardner	791	1/1/1827	1/18/1851
Rice, John Parker	41	6/15/1827	4/25/1832

Reader	Charges	First Charge	Last Charge
Rich, Samuel H.	240	1/13/1832	12/21/1842
Richards, John, Jr.	8	2/1/1827	7/27/1827
Richards, Reuben, Jr.	436	1/14/1834	4/7/1844
Richardson, Albert L.	14	3/30/1850	12/21/1850
Richardson, Jeffrey	23	2/13/1827	12/21/1827
Richardson, Jesse Putnam	101	1/31/1834	8/15/1836
Richardson, William	21	4/8/1848	12/23/1848
Ripley, George	737	4/6/1829	4/9/1841
Robbins, Chandler	161	11/13/1833	2/5/1851
Robbins, Chandler, Jr.	223	2/5/1827	10/21/1833
Robbins, Edward Hutchinson	69	1/10/1833	12/21/1849
Robbins, James Murray	622	1/7/1835	12/21/1850
Robinson, E.	127	2/15/1832	12/13/1836
Robinson, Horatio	124	3/12/1828	11/6/1831
Rogers, Henry Bromfield	241	2/19/1827	10/12/1850
Rogers, Henry D.	32	10/26/1846	3/25/1850
Rogers, William M.	209	9/9/1835	11/25/1850
Rollins, Ebenezer	381	1/19/1833	7/13/1844
Rollins, Ebenezer, Mrs.	128	7/5/1844	1/31/1851
Ropes, William	2	9/5/1850	11/22/1850
Rotch, Benjamin Smith	3	2/10/1847	6/20/1847
Russell, Benjamin	118	6/29/1827	9/8/1828
Russell, LeBaron	26	10/1/1845	12/24/1846
Russell, Samuel Hammond	75	3/24/1845	11/1/1850
Salisbury, Edward Elbridge	23	1/17/1833	6/15/1833
Salisbury, Samuel	90	7/24/1828	12/17/1841
Salter, Richard H.	21	1/31/1850	2/6/1851
Sampson, Reed	17	10/2/1849	6/21/1850
Samson, Edwin	30	8/7/1850	2/8/1851
Sargent, Howard	50	4/28/1831	1/8/1834
Sargent, Ignatius	28	1/21/1849	11/5/1850
Sargent, Lucius Manlius	12	3/15/1849	1/21/1851

Reader	Charges	First Charge	Last Charge
Savage, James	12	1/1/1848	9/10/1851
Sawyer, William	205	1/3/1833	12/24/1836
Sayles, Maria Francreur	20	10/20/1849	1/18/1851
Sayles, Willard	34	12/6/1844	2/12/1849
Scholfield, Arthur	661	2/13/1841	2/1/1851
Searle, George W.	142	2/2/1827	10/24/1851
Sears, David	2	1/1/1827	8/30/1850
Seaver, Benjamin	91	8/7/1844	2/8/1851
Shackford, Charles Chauncy	111	7/22/1841	11/14/1850
Shattuck, George Cheyne	84	1/1/1828	9/6/1848
Shattuck, George Cheyne, Jr.	16	2/4/1847	1/17/1851
Shattuck, Lemuel	160	1/18/1839	10/30/1850
Shaw, Charles Brown	160	1/4/1847	2/11/1851
Shaw, Lemuel	124	2/21/1846	2/14/1851
Shaw, Robert Gould	217	4/16/1833	2/7/1851
Shaw, Robert Gould, Jr.	17	3/11/1850	2/5/1851
Sheafe, Charles C.	8	2/21/1850	4/16/1850
Shelton, Philo Strong	6	10/4/1849	3/29/1850
Sherwin, Thomas	354	12/11/1833	1/9/1844
Shimmin, Charles Franklin	57	3/25/1848	10/12/1850
Shurtleff, Samuel Atwood	348	1/29/1831	12/20/1847
Sigourney, Henry	654	7/4/1827	7/21/1850
Simpson, Michael Hodge	2	4/12/1847	4/12/1847
Skelton, James	8	9/19/1850	1/4/1851
Skinner, Francis	120	4/17/1839	2/6/1851
Smith, Amos	65	1/10/1843	7/10/1848
Smith, Charles Card	41	3/9/1850	1/1/1851
Smith, Henry W.	65	9/1/1849	2/5/1851
Smith, James W.	319	10/1/1844	3/30/1849
Snelling, G.H.	60	2/8/1827	1/13/1830
Sohier, William Davies	12	12/9/1846	1/1/1851
Sprague, Charles	1	1/1/1848	1/1/1850

Reader	Charges	First Charge	Last Charge
Stackpole, J.L.	32	1/21/1843	3/10/1847
Stanton, Francis	36	1/11/1830	12/5/1835
Stanwood, Lemuel	66	4/4/1848	12/14/1850
Starbuck, C.C.	13	2/6/1827	12/20/1827
Stearns, John	154	1/8/1847	1/27/1851
Stevenson, J.G.	95	6/11/1829	2/3/1835
Stimson, Caleb Morton	3	1/1/1827	12/28/1850
Stone, Henry O.	72	3/30/1848	1/30/1851
Stone, William W.	1	4/15/1848	1/1/1850
Storer, D.H.	351	3/30/1839	1/10/1851
Storrow, Charles Storer	12	10/1/1849	11/20/1850
Story, Franklin H., Jr.	33	2/8/1849	9/25/1850
Story, Mrs.	55	1/1/1849	12/3/1850
Story, W.W.	44	1/8/1847	11/10/1847
Strong, Woodbridge	610	2/1/1827	2/11/1851
Sturgis, R.L.	37	1/4/1843	1/24/1844
Sturgis, William	511	2/5/1827	2/12/1850
Sullivan, Richard	348	2/27/1827	12/12/1844
Sullivan, William	133	8/30/1827	12/4/1839
Sumner, Charles	364	3/9/1843	2/11/1851
Swett, Samuel Woodbury	387	2/1/1827	2/11/1851
Tappan, Charles	55	1/17/1829	1/6/1831
Tappan, John	694	1/12/1830	11/16/1850
Taylor, Charles	16	1/16/1834	4/28/1841
Teft, B.F.	4	2/10/1844	4/5/1844
Thacher, Charles	243	2/10/1849	2/3/1851
Thatcher, Meagoun	4	8/8/1850	1/7/1851
Thayer, Andrew E.	62	6/2/1848	5/5/1849
Thayer, John Eliot	11	1/30/1841	12/10/1842
Thomas, Alexander	16	9/29/1842	1/18/1844
Thorndike, Charles	7	1/1/1831	3/2/1831
Thorndike, Israel, Jr.	265	1/1/1827	2/6/1837

Reader	Charges	First Charge	Last Charge
Ticknor, George	242	1/29/1827	12/31/1850
Tilden, Joseph	438	1/1/1827	10/29/1850
Timmins, Henry	11	1/1/1847	1/26/1850
Torrey, Charles	158	1/3/1827	12/4/1850
Torrey, Samuel	658	2/21/1827	2/5/1851
Tower, David B.	44	7/1/1837	12/29/1838
Towne, Joseph	78	1/13/1837	1/26/1846
Townsend, Solomon Davis	483	6/11/1831	1/11/1851
Treadwell, Daniel	149	2/5/1827	9/22/1837
Tuckerman, Edward	716	1/11/1831	1/11/1842
Tuckerman, Gustavus	50	2/23/1836	5/28/1837
Tuckerman, Henry Harris	1326	1/3/1832	12/2/1841
Tuckerman, Joseph	649	1/9/1828	4/29/1840
Upham, Jabez Baxter	67	4/7/1848	12/28/1850
Upton, George Bruce	16	1/14/1850	2/3/1851
Wainwright, Jonathan M.	59	8/9/1832	9/7/1835
Waldo, Henry S.	2	1/18/1851	2/12/1851
Waldo, Henry S., Jr.	6	3/7/1845	12/29/1845
Waldron, Samuel K.	24	7/5/1836	12/27/1837
Wales, Thomas Beale	426	2/25/1837	6/18/1847
Walker, Charles	11	8/18/1828	12/3/1828
Walker, William Johnson	21	4/24/1828	12/27/1828
Walley, Samuel H.	3	6/21/1827	6/21/1827
Ward, Thomas Wren	888	1/31/1827	9/18/1850
Ware, Charles Eliot	33	10/30/1849	12/27/1850
Ware, John	1021	1/30/1827	1/22/1851
Warren, A., Mrs.	225	2/15/1827	6/28/1833
Warren, Charles H.	50	1/9/1847	10/31/1850
Warren, Edward	911	7/20/1833	1/7/1851
Warren, George Willis	43	4/14/1838	12/7/1838
Warren, J.W.	30	3/31/1843	3/29/1844
Warren, John Collins	757	2/10/1827	2/11/1851

Reader	Charges	First Charge	Last Charge
Warren, Jonathan Mason	164	1/11/1844	1/8/1851
Waterston, Robert Cassie	187	2/6/1827	11/18/1843
Watson, John Lee	83	2/14/1837	12/20/1837
Webb, Seth, Jr.	50	1/18/1850	2/1/1851
Webster, Daniel	12	1/17/1828	4/9/1834
Webster, Fletcher	5	4/5/1845	11/1/1849
Welch, John Hunt	36	11/19/1849	2/11/1851
Welles, Arnold	141	2/18/1837	7/6/1842
Welles, Benjamin	100	1/3/1838	11/22/1842
Wells, William	15	4/19/1834	11/14/1834
Weston, Alden Bradford	326	2/18/1833	12/20/1850
Wheelwright, William Wilson	5	10/10/1827	1/9/1828
Whipple, Charles K.	199	3/11/1847	2/8/1851
Whipple, Edwin Percy	119	1/6/1849	2/8/1851
Whipple, John A.	2	5/2/1851	5/23/1851
White, William Orne	14	4/13/1850	2/6/1851
Whiting, Nathaniel	7	1/1/1849	3/5/1849
Whiting, William	31	2/23/1850	2/8/1851
Whitmore, Charles John	14	7/13/1850	1/29/1851
Whitney, Benjamin Duick	251	1/8/1831	2/5/1851
Whitney, Joseph	59	1/29/1845	12/21/1850
Whitney, Josiah Dwight	7	3/6/1850	12/8/1850
Whitney, Moses, Jr.	271	1/18/1828	12/25/1839
Whitney, Warren Jacob	11	2/25/1843	12/23/1843
Whitney, William Fiske	103	4/1/1845	8/28/1850
Wigglesworth, Edward	665	4/1/1831	11/8/1850
Wigglesworth, Thomas	312	1/31/1827	1/1/1851
Wild, Charles	127	2/26/1833	10/14/1837
Willard, Solomon	88	2/14/1827	7/19/1844
Williams, Henry Willard	94	11/20/1849	2/4/1851
Williams, John Davis	2	1/1/1827	8/18/1846
Williams, Moses Blake	135	10/2/1845	1/24/1851

Reader	Charges	First Charge	Last Charge
Williams, Samuel King	12	3/16/1848	4/19/1848
Williams, Thomas	25	1/10/1831	4/10/1832
Williams, Timothy	165	2/6/1827	11/28/1840
Wills, William, Jr.	53	1/23/1835	4/21/1838
Wing, Benjamin Franklin	114	2/24/1846	2/4/1851
Winslow, Isaac	587	2/10/1827	2/15/1851
Wright, Hartley Hezekiah	106	7/28/1838	1/25/1840
Wright, John H.	10	11/16/1847	8/19/1848
Wyman, Jeffers	67	6/7/1844	2/8/1851
Wyman, Rufus	200	1/31/1827	8/10/1844
Young, Alexander	454	2/3/1829	10/5/1842
Young, Ammi Burnham	2	1/22/1842	1/22/1842
Young, Edward J.	18	9/14/1848	9/11/1849

Appendix C

Readers in Experimental Philosophy

Reader	Titles	Charges
Abbot, Samuel L.	1	1
Abbott, Jacob	6	6
Adams, Benjamin	2	2
Adams, Charles Francis	3	3
Adams, Francis M.	2	2
Adams, L.B.	1	1
Adams, Zabdiel B.	3	3
Alger, Cyrus	14	17
Alger, Francis	8	10
Amory, Charles	2	2
Amory, Thomas Coffin, Jr.	7	9
Amory, William	3	4
Andrews, Caleb	5	6
Andrews, James	2	2
Andrews, William Turell	11	33
Appleton, Nathan	5	6
Appleton, Samuel	1	1

Reader	Titles	Charges
Atkinson, Edward	6	8
Atkinson, William Parsons	3	3
Atwood, Charles	8	9
Austin, James Trecothick	4	4
Austin, Samuel, Jr.	6	7
Baker, A.B.	1	1
Bancroft, George	1	1
Barnard, Charles	14	15
Barrell, Samuel B.	2	3
Barrett, Samuel	5	5
Bartlett, Sidney	1	1
Bartol, Cyrus Augustus	7	8
Bass, Seth	1	1
Bates, George	27	30
Beck, Charles	1	1
Belknap, Andrew Eliot	1	1
Belknap, John	29	46
Bell, Joseph	3	3
Bethune, John McLean	2	2
Bigelow, Erastus Brigham	1	1
Bigelow, Jacob	45	86
Binney, Amos	1	1
Binney, Amos, Mrs.	1	1
Bird, John A.	2	2
Blagden, George W.	2	7
Blake, Edward	1	3
Bond, George	5	6
Boott, Francis	4	4
Boott, John W.	17	37
Boott, Kirk	9	11
Borland, John Nelson	1	1
Bowditch, Henry Ingersoll	4	4

Reader	Titles	Charges
Bowditch, Jonathan Ingersoll	3	3
Bowditch, Nathaniel	5	13
Bowditch, Nathaniel Ingersoll	6	8
Boyden, Uriah Atherton	11	21
Bradbury, Charles	5	5
Bradford, Thomas G.	4	7
Bradlee, Josiah	3	3
Brewer, Thomas Mayo	1	1
Brimmer, George W.	5	7
Brimmer, Martin	3	3
Brooks, Benjamin	2	2
Brooks, Edward	2	2
Brooks, Francis Augustus	1	1
Brooks, Gorham	7	8
Brooks, Peter Chardon	3	3
Brooks, William H.	3	4
Bullard, William Story	4	5
Bumstead, John	2	2
Burley, Susan	19	28
Burroughs, George	3	3
Butler, Clement M.	1	1
Cabot, Samuel	6	6
Cabot, Samuel, Jr.	3	7
Capen, Nahum	21	29
Cary, Thomas Greaves	2	2
Chandler, Abiel	2	3
Chandler, Gardiner Leonard	3	3
Chandler, Gardiner Leonard, Jr.	3	3
Channing, Walter	4	4
Channing, William Ellery	2	3
Chapman, Jonathan	1	1
Chapman, O.G.	1	1

Reader	Titles	Charges
Chase, Theodore	1	1
Choate, Rufus	9	9
Clarke, James Freeman	13	13
Clinch, Joseph H.	1	1
Cobb, Richard	3	3
Codman, Henry	28	39
Coffin, George Washington	1	1
Coffin, John Gorham	2	2
Coffin, Joy	16	20
Coffin, Thomas M.	3	6
Colburn, Warren	8	10
Cooke, Josiah Parsons, Jr.	22	31
Cooke, Josiah Parsons, Sr.	24	28
Coolidge, Joseph, Jr.	3	3
Cordis, Thomas	1	1
Cotting, B.E.	1	1
Courties, Ambrose S.	3	3
Cruft, Edward	20	22
Curtis, Benjamin Robbins	4	5
Curtis, Caleb	1	2
Curtis, Charles Pelham	13	17
Curtis, Nathaniel, Jr.	1	1
Curtis, Thomas Buckminster	3	3
Cushing, Caleb	2	2
Cushing, Thomas Parkman	1	1
Dana, Richard Henry, Jr.	2	2
Danforth, Bowers	1	1
Darling, Benjamin	2	3
Davis, Edward Gardiner	13	13
Davis, Francis	1	2
Davis, Isaac P.	12	15
Davis, John	9	15

Reader	Titles	Charges
Davis, John Brazer	9	10
Davis, Jonathan	1	1
Davis, Thomas	1	1
Dearborn, Henry Alexander Scammell	4	7
Degrand, Peter P.F.	1	1
Dexter, Franklin	36	70
Dexter, George Minot	8	8
Dillaway, Charles Knapp	6	7
Dixwell, Epes Sargent	3	3
Dixwell, John James	12	13
Dodge, Pickering	1	1
Dorr, Charles Hazen	1	1
Dorr, John	1	1
Dorr, Samuel	2	4
Downer, Samuel, Jr.	5	5
Dwight, Edmund	7	7
Eaton, Asa	7	9
Eddy, Robert Henry	17	49
Edwards, B.B.	3	3
Eldridge, Oliver	1	1
Eliot, Samuel Atkins	5	9
Ellis, David	4	4
Emerson, George Barrell	45	78
Emerson, Ralph Waldo	12	15
Everett, Alexander H.	6	6
Fairbanks, Charles B.	1	1
Fairbanks, Gerry	1	1
Fairbanks, Stephen	7	8
Faxon, Nathaniel	2	2
Felt, Joseph Barlow	3	4
Flagg, Josiah Foster	6	6
Fletcher, Richard	7	7

Reader	Titles	Charges
Folsom, Charles	36	46
Forbes, Robert Bennet	9	12
Foster, John	3	3
Foster, Thomas	5	7
Foster, William	1	1
Fox, Thomas B.	1	1
Francis, Ebenezer	2	2
Freeman, James	5	6
Freeman, James, Jr.	2	2
Frothingham, Nathaniel Langdon	10	13
Gardiner, Chandler L., Jr.	6	11
Gardiner, William Howard	3	3
Gay, Martin	10	21
Gibbs, William	1	1
Goodrich, Samuel Grisold	25	50
Gorham, Benjamin	1	1
Gorham, John	9	11
Gould, Augustus Addison	18	25
Gragg, Washington Parker	3	3
Gray, Francis Calley	15	16
Gray, Horace	26	46
Gray, Horace, Jr.	5	6
Gray, John Chipman	2	2
Greene, Benjamin Daniel	4	4
Greene, Gardiner	2	2
Greene, Gardiner, Mrs.	4	5
Greene, Nathaniel	6	8
Greenleaf, Ezekiel Price	3	4
Greenough, Alfred	7	7
Grigg, William	27	34
Guardenier, John	19	22
Guild, Benjamin	5	5

Reader	Titles	Charges
Hale, Enoch	4	5
Hale, Nathan	18	24
Hall, James	3	3
Hall, Joseph	30	64
Hall, Nathaniel	1	1
Hall, Thomas Bartlett	1	1
Hammond, Daniel	4	7
Hancock, John	7	8
Hancock, John, Jr.	1	1
Hancock, John, Sr.	1	1
Harris, Thaddeus Mason	18	32
Hastings, Daniel	7	7
Hayes, Francis Brown	2	3
Hayward, George	10	16
Hayward, James	3	4
Hayward, Joshua H.	2	3
Healy, Mark	1	1
Heard, John, Jr.	42	54
Henderson, A.A.	4	4
Henshaw, David	8	11
Henshaw, Samuel	1	1
Hesmer, Z.	1	1
Higginson, Francis	10	13
Higginson, Henry	11	12
Higginson, James Perkins	1	2
Higginson, Waldo	2	2
Hillard, George Stillman	3	3
Himes, Joshua Vaughn	1	1
Hobbes, Prentiss	3	3
Hoit, A.G.	1	1
Homans, I. Smith	3	3
Homer, Joseph Warren	1	1

Reader	Titles	Charges
Hooper, Henry N.	18	21
Hooper, Robert William	7	8
Hooper, Samuel	12	24
Howe, Joseph N., Jr.	1	2
Howe, Samuel Gridley	15	19
Hunt, E.B.	6	6
Inches, Henderson	19	23
Ingalls, William	2	2
Jackson, Charles	23	27
Jackson, Lydia	10	11
Jackson, Patrick Tracy	25	34
Jenks, John H.	1	1
Jenks, William	22	56
Jones, Anna P.	6	25
Jones, I., Mrs.	1	1
Jones, John C.	1	1
Joy, Benjamin	2	2
Kettell, Samuel	5	7
Kirkland, John Thornton	1	1
Knowles, Seth	3	3
Lamb, Thomas	4	4
Lawrence, Abbot	6	7
Lawrence, Amos Adams	1	1
Lawrence, William Richards	23	37
Lee, Henry	10	11
Lee, Henry, Jr.	1	1
Lee, Thomas, Jr.	6	7
Lieber, Francis	5	5
Livermore, George	1	1
Lloyd, James	1	1
Lodge, Giles Henry	2	2
Loring, Benjamin	1	1

Reader	Titles	Charges
Loring, Caleb	2	2
Loring, Charles Greely	2	3
Loring, Edward Greely	1	1
Loring, Elijah	1	1
Loring, Josiah	3	4
Lovering, Joseph Swain	1	1
Lowell, Francis Cabot	6	10
Lowell, John	1	1
Lowell, John Amory	8	11
Lowell, John, Jr.	3	3
Lowell, John, Sr.	1	1
Lunt, William P.	2	2
Lyman, Theodore	5	17
Lyman, William	28	68
Martin, Enoch	26	66
Mason, Robert Means	3	3
Mason, William Powell	2	2
McKean, Joseph W.	3	7
Merrill, Amos B.	1	1
Miles, Solomon Pierson	24	34
Moody, Paul	1	1
Motley, Thomas	2	3
Mott, J.M.	8	8
Nichols, Benjamin Ropes	29	48
Odin, John	3	4
Oliver, Eldridge	3	3
Osgood, David	11	18
Otis, George Washington, Jr.	16	20
Otis, Harrison Gray	3	3
Page, Henry Augustus	6	6
Parker, Daniel Pinckney	4	5
Parker, John, Jr.	10	13

Reader	Titles	Charges
Parker, Samuel P.	4	4
Parker, Theodore	4	4
Parker, William A.	1	3
Parkman, Francis	19	32
Parkman, George	14	14
Parkman, Samuel	3	3
Parris, Alexander	30	77
Parrott, William Pearce	3	3
Parsons, Theophilus	16	29
Parsons, William	17	19
Payne, Edward William	1	1
Peabody, Ephraim	7	9
Perkins, Stephen Higginson	17	22
Perkins, Thomas Handasyd	6	6
Perkins, William Powell	2	2
Phillips, Jonathan	23	26
Pickering, Octavius	5	5
Porter, Jonathan	3	5
Pratt, William	10	16
Prescott, William Hickling	9	12
Putnam, Jesse P.	13	14
Quincy, Edmund	3	3
Quincy, Josiah	9	11
Quincy, Josiah, Jr.	25	32
Randall, Elizabeth	1	1
Randall, John	15	29
Reynolds, Edward, Jr.	16	29
Rice, Henry	4	4
Rice, Henry Gardner	6	8
Rice, John Parker	2	2
Rich, Samuel H.	10	18
Richards, Reuben, Jr.	4	6

Reader	Titles	Charges
Richardson, Jesse Putnam	3	3
Ripley, George	15	20
Robbins, Chandler	4	4
Robbins, Chandler, Jr.	5	6
Robbins, James Murray	12	13
Rogers, Henry D.	9	10
Rogers, William M.	2	2
Rollins, Ebenezer, Mrs.	2	2
Russell, Benjamin	1	1
Sargent, Howard	3	3
Savage, James	2	2
Sawyer, William	1	1
Sayles, Maria Francreur	1	4
Scholfield, Arthur	16	33
Searle, George W.	2	2
Shackford, Charles Chauncy	2	2
Shattuck, George Cheyne	1	2
Shattuck, George Cheyne, Jr.	1	1
Shattuck, Lemuel	2	4
Shaw, Charles Brown	2	4
Shaw, Lemuel	5	5
Shaw, Robert Gould	2	2
Sheafe, Charles C.	3	3
Shelton, Philo Strong	2	2
Sherwin, Thomas	33	44
Shurtleff, Samuel Atwood	13	14
Skinner, Francis	7	7
Smith, Henry W.	2	2
Snelling, G.H.	2	3
Starbuck, C.C.	8	11
Stearns, John	1	1
Stevenson, J.G.	4	5

Reader	Titles	Charges
Stone, William W.	1	1
Storer, D.H.	4	5
Storrow, Charles Storer	1	1
Strong, Woodbridge	8	11
Sturgis, William	3	5
Sullivan, Richard	6	11
Sullivan, William	1	1
Swett, Samuel Woodbury	10	11
Tappan, John	15	16
Thayer, John Eliot	3	3
Thorndike, Israel, Jr.	5	6
Ticknor, George	2	2
Tilden, Joseph	4	6
Torrey, Samuel	1	1
Tower, David B.	3	3
Townsend, Solomon Davis	5	5
Treadwell, Daniel	19	23
Tuckerman, Edward	45	82
Tuckerman, Henry Harris	2	2
Tuckerman, Joseph	9	11
Wainwright, Jonathan M.	1	2
Waldron, Samuel K.	2	2
Wales, Thomas Beale	2	2
Ward, Thomas Wren	5	10
Ware, Charles Eliot	1	2
Ware, John	12	20
Warren, A., Mrs.	10	24
Warren, Edward	15	18
Warren, John Collins	15	24
Warren, Jonathan Mason	6	6
Waterston, Robert Cassie	1	1
Webb, Seth, Jr.	1	1

Reader	Titles	Charges
Weston, Alden Bradford	12	18
Whipple, Charles K.	4	8
Whipple, Edwin Percy	1	1
Whiting, William	1	2
Whitney, Benjamin Duick	2	2
Whitney, Moses, Jr.	6	6
Wigglesworth, Edward	19	30
Wigglesworth, Thomas	2	2
Wild, Charles	1	1
Willard, Solomon	1	2
Williams, Timothy	1	1
Wills, William, Jr.	3	3
Wing, Benjamin Franklin	3	5
Winslow, Isaac	7	11
Wright, John H.	1	1
Wyman, Jeffers	7	8
Wyman, Rufus	21	35
Young, Alexander	6	6
Young, Edward J.	7	7

Appendix D

Top Readers and Top Titles by Subject Matter Category

D.1 Top Five Readers in Each Subject Matter Category

The number under **Readers** for each subject matter category is the total number of readers of that category.

	Readers	Charges
PERIODICAL	121	494
Martin, Enoch		35
Hall, Joseph		31
Eddy, Robert Henry		30
Parris, Alexander		28
Dexter, Franklin		25
MECHANICS	136	481
Parris, Alexander		42
Bigelow, Jacob		40
Jenks, William		28
Nichols, Benjamin Ropes		19
Lyman, William		14

	Readers	Charges
CHEMISTRY	130	459
Dexter, Franklin		48
Tuckerman, Edward		21
Belknap, John		20
Cooke, Sr., Josiah		18
Hall, Joseph		17
GEOLOGY	172	445
Martin, Enoch		18
Codman, Henry		16
Randall, John		9
Belknap, John		8
Parsons, William		8
BOTANY	106	335
Boott, John W.		26
Emerson, George B.		25
Tuckerman, Edward		25
Lyman, Theodore		15
Grigg, William		11
ZOOLOGY	106	323
Andrews, William Turell		27
Reynolds, Jr., Edward		16
Jones, Anna P.		14
Warren, Mrs. A.		12
Emerson, George B.		12
ASTRONOMY	124	275
Miles, Solomon		17
Goodrich, Samuel		15
Cooke, Jr., Josiah P.		9
Jackson, Patrick T.		7
Bowditch, Nathaniel		7

	Readers	Charges
MATHEMATICS	82	227
Sherwin, Thomas		26
Heard, Jr., John		13
Miles, Solomon		10
Guardenier, John		8
Lawrence, William Richards		7
ENTOMOLOGY	89	222
Parsons, Theophilus		13
Harris, Thaddeus M.		12
Capen, Nahum		10
Osgood, David		9
Codman, Henry		9
HYDRAULICS	64	161
Parris, Alexander		21
Lyman, William		19
Eddy, Robert H.		11
Gray, Horace		8
Forbes, Robert Bennet		6
PYRONOMICS	65	130
Wyman, Rufus		12
Gray, Horace		9
Barnard, Charles		5
Parris, Alexander		5
Gray, Francis Calley		5
MINERALOGY	59	123
Parkman, Francis		8
Codman, Henry		7
Tuckerman, Edward		7
Jenks, William		6
Alger, Cyrus		5

	Readers	Charges
ORNITHOLOGY	67	122
Goodrich, Samuel		9
Whipple, Charles K.		7
Cabot, Jr., Samuel		6
Hooper, Samuel		6
Reynolds, Jr., Edward		4
CONCHOLOGY	44	114
Tuckerman, Edward		17
Pratt, William		10
Jones, Anna P.		9
Parsons, Theophilus		7
Inches, Henderson		6
PATENTS	59	108
Eddy, Robert H.		23
Goodrich, Samuel		5
Folsom, Charles		4
Wigglesworth, Edward		4
Lyman, William		4
SILLIMAN'S JOURNAL	38	98
Hall, Joseph		18
Martin, Enoch		17
Heard, Jr., John		7
Lyman, William		7
Folsom, Charles		4
FRANKLIN JOURNAL	38	92
Hall, Joseph		12
Bigelow, Jacob		12
Lyman, William		7
Jenks, William		7
Gray, Horace		6

	Readers	Charges
Bigelow's *Technology*	34	46
Randall, John		5
Rice, Henry Gardner		2
Howe, Samuel Gridley		2
Brooks, William H.		2
Warren, John Collins		2

D.2   Top Five Titles in Each Subject Matter Category

The number under **Short Title** for each subject matter category is the total number of titles in that category.

Author	Short Title	Readers	Charges
Periodicals	13	121	494
Robertson	*London Mechanics Magazine*	37	103
Silliman	*American Journal of Science*	38	98
Jones	*Franklin Journal*	38	92
Newton	*Repertory of Patent Inventions*	27	60
Lavoisier	*Annales de Chimie*	20	52
Mechanics	90	136	481
Robertson	*London Mechanics Magazine*	37	103
Arnott	*Elements of Physics*	21	26
Wood	*Practical Treatise on Rail Roads*	14	24
Moseley	*Treatise on Mechanics*	12	17
Ure	*Cotton Manufacture*	7	16
Chemistry	52	130	459
Liebig	*Organic Chemistry in Agriculture*	35	53
Lavoisier	*Annales de Chimie*	20	52
Faraday	*Chemical Manipulation*	20	37
Turner	*Elements of Chemistry*	23	36
Webster	*Manual of Chemistry*	18	24

Author	Short Title	Readers	Charges
GEOLOGY	45	172	445
Lyell	*Geology*	54	88
Bakewell	*Introduction to Geology*	28	33
Cuvier	*Essay on the Theory of the Earth*	25	32
Buckland	*Geology and Mineralogy Considered*	21	26
Greenough	*Transactions Geological Society*	9	26
BOTANY	71	106	335
Curtis	*Botanical Magazine*	13	50
Loudon	*Encyclopedia of Plants*	14	29
Bigelow	*American Medical Botany*	17	25
Sowerby	*English Botany*	7	22
Bigelow	*Florula Bostoniensis*	15	19
ZOOLOGY	40	106	323
Cuvier	*Animal Kingdom*	33	67
Shaw	*General Zoology*	21	44
Cuvier	*Histoire Naturelle des Poissons*	8	33
Lamarck	*Animaux Sans Vertebres*	14	30
Bingley	*Animal Biography*	12	20
ASTRONOMY	44	124	275
Humboldt	*Cosmos*	54	72
Whewell	*Astronomy and General Physics*	18	26
Nichol	*Architecture of the Heavens*	20	24
Laplace	*System of the World*	13	17
Laplace	*Mécanique Céleste*	6	13

Author	Short Title	Readers	Charges
MATHEMATICS	86	82	227
Priestley	Familiar Introduction to Perspective	8	11
Lacroix	Treatise on the Differential and Integral Calculus	6	11
Euler	Letters to a German Princess	7	9
Hutton	Mathematical Tables	4	7
Simson	Elements of Euclid	6	6
ENTOMOLOGY	25	89	222
Kirby	Introduction to Entomology	33	61
Huber	Natural History of Bees	18	21
Huber	Natural History of Ants	19	20
Major	Insects on Fruit Trees	10	14
Harris	Insects of Massachusetts	14	14
HYDRAULICS	24	64	161
Tredgold	Steam Engine Investigation of its Principles	14	24
Lardner	The Steam Engine	21	24
Farey	Treatise on the Steam Engine	8	23
Matthews	Hydraulia	12	19
Stuart	Descriptive History of the Steam Engine	9	13
PYRONOMICS	10	65	130
Tredgold	Principles of Warming and Ventilating	25	39
Arnott	On Warming and Ventilating	14	18
Wyman	Treatise on Ventilation	14	16
Leslie	Experimental Inquiry into the Propagation of Heat	9	13
Hood	Treatise on Warming Buildings	9	10

Author	Short Title	Readers	Charges
MINERALOGY	18	59	123
Cleaveland	*Treatise on Mineralogy and Geology*	21	34
Dana	*Mineralogy and Geology of Boston*	13	14
Jacob	*Precious Metals*	11	13
Phillips	*Elementary Introduction to Mineralogy*	10	11
Bakewell	*Introduction to Mineralogy*	7	8
ORNITHOLOGY	18	67	122
Nuttall	*Manual of Ornithology*	25	30
Audubon	*Ornithological Biography*	19	23
Wilson	*American Ornithology*	10	13
Buffon	*Natural History of Birds*	11	12
Swainson	*Classification of Birds*	5	8
CONCHOLOGY	12	44	114
Burrow	*Elements of Conchology*	15	22
Wood	*General Conchology*	15	20
Dillwyne	*Descriptive Catalogue of Shells*	7	19
Mawe	*Linnaean System of Conchology*	14	18
Brookes	*Study of Conchology*	7	9
PATENTS	6	59	108
Newton	*Repertory of Patent Inventions*	27	60
Beckmann	*History of Inventions*	26	38
Worcester	*Century of Innovations*	4	4
Ellsworth	*Digest of United States Patents*	2	2
Treuttel	*Archives des Inventions*	2	2

Appendix E

Patents of Readers at the Boston Athenæum (1790–1850)

Twenty-four members of the Boston Athenæum whose names appear in the first four volumes of the Athenæum's books borrowed register were granted as a group ninety US patents between 1790 and 1850. Table E.1 is a tabulation of number of patents granted to each member while Table E.1 lists the issue date and title of each patent for each member.

As is always the case in such matters, associations between a name on a patent and a name in the books borrowed register are not of uniform certainty. The Cyrus Alger who was granted patents for metallurgy innovations was almost certainly the Athenæum member Cyrus Alger whose name is found in volumes I, II, and III of the books borrowed register. The Augustus A. Gould whose book borrowing is recorded in all four volumes is almost certainly the Augustus Addison Gould whose name appears on the famous ether patent. Whether or not book borrower Charles Homer is the Charles S. Homer who invented an "Improvement in the Manner of Constructing Garden Hoes" or whether it was Athenæum member Michael Simpson who invented a "Strap for Pantaloons" are questions worthy of review.

In addition to name matching, account was taken of the recorded residence of the patent holder and concurrence of what might be known about an Athenæum member with the subject matter of the patent. All of the patents associated with the readers were granted to residents of Massachusetts. None of the patents were for an improved churn. To give a sense of the fuzziness of the line between attribution or not, Table E.2 lists patents for which the match between patent data and individuals whose names are found in the books borrowed was deemed insufficient for inclusion in the counts.

The patents whose titles are in lower-case italics are X patents which have not been recovered. The titles of these lost patents are taken from *A List of patents granted by the United States, for the encouragement of arts and sciences, alphabetically arranged from 1790 to 1828* [63], *List of Patents for Inventions and Designs Issued by the United States from 1790 to 1847* [49] and *A List of Patents Granted by the United States from April 10, 1790, to December 31, 1836 &c.* [64]. All other titles are taken from the patents as retrieved from the United States Patent and Trademark Office.

Reader	Patents	Reader	Patents
Alger, Cyrus	5	Gould, Augustus A.	1
Bigelow, Erastus B.	20	Hale, Nathan	2
Boyden, Uriah A.	3	Higginson, Henry	1
Chandler, Abiel	1	Homer, Charles S.	1
Curtis, Caleb	1	Hurd, Joseph Jr.	11
Dexter, George M.	1	Jackson, Patrick T.	1
Jackson, Charles T.	1	Lyman, William	1
Eddy, Robert H.	1	Moody, Paul	10
Flagg, Josiah F.	2	Savage, James S.	1
Forbes, Robert B.	1	Robinson, Enoch	6
Foster, Leonard	6	Treadwell, Daniel	10
Goodrich, Samuel	1	Whipple, John A.	2

Table E.1: Patent Counts for Athenæum Readers[1]

---

[1] The counts do not include reissued patents.

E.1 Boston Athenæum Patent Holders and Their Patents

Issue Date	Number	Title
CYRUS ALGER		
3/30/1811	1,483X	Casting Iron Rolls
5/18/1814	*2,136X*	*casting iron hinges*
7/1/1836	*9,817X*	*sheaves for blocks*
5/30/1837	208	Improvement in the Art of Manufacturing Cast-Iron Cannon
8/3/1838	869	Improved Manufacture of Plows of Malleable Cast-Iron
ERASTUS BRIGHAM BIGELOW		
1/27/1835	*8,620X*	*setting awls, drills, etc.*
4/20/1837	169	Power-Loom for Weaving Coach-Lace and Other Similar Fabrics
1/6/1838	546	Loom for Weaving Knotted Counterpanes and Other Fabrics in which the Woof is Raised from the Surface
4/24/1840	1,561	Improvement in the Mode of Constructing the Power-Loom so as to Adapt it to the Weaving of Figured Counterpanes and Other Articles
5/16/1842	2,625	Loom for Weaving Carpets and Other Figured Fabrics
5/26/1842	2,639	Manner of Constructing Looms for Weaving Carpets and Other Similarly Wrought Fabrics
5/30/1842	2,653	Manner of Mounting Looms for Weaving Counterpanes and Other Articles
7/28/1842	2,741	Improvement in Power-Looms for Weaving Counterpanes, &c.
8/2/1842	2,744	Power-Loom for Weaving Counterpanes, &c.
2/24/1845	3,925	Improvement in Jaw-Temples for Weaving-Looms
2/24/1845	3,926	Flier of Speeders

Issue Date	Number	Title
3/12/1845	3,948	Apparatus for Regulating the Tension of Warp in Looms
4/10/1845	3,987	Improvement in Power-Looms for Weaving Plaids, &c.
2/18/1846	4,696	Power-Loom
9/26/1846	RE87	Reissue of 169
3/20/1847	5,020	Brussels Loom
9/5/1848	5,754	Apparatus for Stretching and Drying Cloth
3/10/1849	6,153	Loom for Weaving Carpets, &c.
3/13/1849	6,186	Loom for Weaving Brussels Carpeting, &c.
9/11/1849	RE143	Reissue of 2,625
9/11/1849	RE144	Reissue of 4,696
10/23/1849	6,806	Jacquard Loom
10/9/1849	RE147	Reissue of 6,153
11/20/1849	RE150	Reissue of 6,186
9/24/1850	7,660	Loom for Weaving Tapestry and Brussels Carpet

JOHN W. BOOTT

12/29/1826	4,624X	construction of fire places, furnaces, etc.[2]

URIAH ATHERTON BOYDEN

4/17/1847	5,068	Hanging Shaft of Water-Wheels, &c.
5/1/1847	5,090	Improvement in Diffusers for Water-Wheels
6/5/1847	5,144	Improvement in Water-Wheels

ABIEL CHANDLER

7/13/1832	7,165X	blocks, cam and spring for fastening stereotype plates

CALEB CURTIS

4/20/1831	6,496X	sheaves of blocks

---

[2] With William Lyman

Issue Date	Number	Title
George M. Dexter		
12/1/1840	1,875	Improvement in the Mode of Heating Buildings by Means of an Apparatus Consisting of Tubes for the Circulation of Hot Water Arranged in an Air-Chamber Adapted to the Same
Robert H. Eddy		
5/10/1844	3,582	Lamp-Cap
Josiah Flagg, Sr. & Jr.		
5/17/1810	1,307X	*elastic lamp reflector or elastic tubes for Argand's lamps*
8/7/1849	6,637	Locomotive Spark-Arrester and Smoke-Conductor
Robert B. Forbes		
12/23/1849	6,815	Apparatus for Distilling Sea-Water[3]
Leonard Foster		
8/20/1813	1,989X	*improvement in the lamp, consisting of a double reflector to one blaze*
2/16/1824	3,817X	windlass bedstead
6/25/1824	3,898X	*windlass bedstead*
2/25/1831	6,396X	fastening mortise door
6/27/1838	811	Mortise Latch for Doors
8/28/1841	2,231	Manner of Constructing Mortise-Latch Door-Fasteners
Samuel Goodrich		
10/11/1828	5,243X	*stereotype blocks*
Augustus A. Gould		
11/13/1847	5,365	Apparatus for Inhaling Ether, &c.

---

[3] With John Ericsson

Issue Date	Number	Title
NATHAN HALE		
11/20/1829	*5,718X*	*improved stereotype plate*
2/15/1833	*7,427X*	*stereotype block*
HENRY HIGGINSON		
4/4/1838	673	Improvement in the Mode of Building Ships and Other Vessels
CHARLES S. HOMER		
5/8/1840	1,593	Improvement in the Manner of Constructing Garden-Hoes
JOSEPH HURD, JR.		
11/10/1829	5,703X	Cook Stove
11/10/1829	5,704X	Boiler, Constructing
11/10/1829	5,705X	Furnace With Reflector
11/10/1829	5,706X	Hot-Air Furnace
1/23/1830	5,792X	Cloth Drying Machine
6/23/1838	806	Reservoir Cooking Stove
7/26/1838	855	Improved Process of Manufacturing Sugar from Beets
7/26/1838	856	Machine for Cutting Beet-Roots for the Manufacture of Sugar
4/24/1841	2,063	Apparatus to be Attached to Chimneys to Increase the Draft and Prevent their Smoking
10/3/1844	3,772	Machine for Separating Liquids from Sugar
12/12/1844	3,854	Cap for Regulating the Draft of Chimneys
CHARLES T. JACKSON		
11/12/1846	4,848	Improvement in Surgical Operations[4]
PATRICK TRACY JACKSON		
2/23/1815	*2,271X*	*looms*[5]

[4] With W.T.G. Morton
[5] With F.C. Lovel

Issue Date	Number	Title
William Lyman		
12/29/1826	4,624X	*construction of fire places, furnaces, etc.*[6]
Paul Moody		
3/9/1816	2,451X	*winding yarn from bobbins or spools*
1/17/1818	2,900X	*machine for sizing and dressing cloth*
4/3/1819	3,091X	*double speeder for roping cotton*
5/6/1819	3,103X	*spinning frame for cotton*
12/30/1820	3,275X	*double speeder*
1/19/1821	3,282X	*machine for roping cotton*
2/19/1821	3,295X	*machine for roping and spinning cotton*
2/19/1821	3,296X	*machine for roping cotton, called the double speeder*
2/19/1821	3,297X	*spinning frames for cotton*
7/8/1834	8,296X	*straw, cutting hay, &c., horizontal*
Enoch Robinson		
10/20/1836	65	Ferrule-Knob for Doors &c[7]
12/2/1836	98	Door, Commode, &c., Knob[7]
10/20/1837	434	Glass Knobs
9/11/1841	2,248	Window Fastener
2/28/1842	2,473	Windlass
1/10/1843	2,904	Door Knob
James S. Savage		
2/15/1838	607	Boom-Derrick

[6] With John W. Boott
[7] With J.H. Lord

Issue Date	Number	Title
DANIEL TREADWELL		
8/8/1817	2,823X	*improvement in their screw machine*[8]
3/2/1826	4,348X	Printing Press, Power
10/11/1831	6,794X	Spinning Fibrous Material
2/3/1834	7,995X	Spinning Machine
2/5/1834	7,997X	Improvements in the Art of Rope Making Called an Iron Tail
8/18/1834	8,366X	Hatcheling Flax and Hemp
8/18/1834	8,367X	*machine for tarring rope yarns*
3/31/1840	1,534	Condensing Apparatus of Steam-Engines
2/12/1845	3,906	Method of Making Cannon of Wrought-Iron or Wrought-Iron and Steel
6/20/1846	4,589	Machinery for Welding and Forming Wrought-Iron Cannon
JOHN A. WHIPPLE		
1/23/1849	6,056	Improvement in Taking Daguerreotype-Pictures
6/25/1850	7,458	Improvement in Producing Photographic Plates upon Transparent Media

[8] With Phinehas Dow

E.2 Patent Holders who may have been Athenæum Readers

Name on Patent	City	Patent Number	Possible Match
Benjamin Franklin Adams	Boston	7782X	Benjamin Adams
	New Bedford	4,574	
Charles Adams	Boston	4,040X	Charles Francis Adams
		6,487X	Charles Frederick Adams
William Blake	Boston	7,559X	William Blake
		4,141	
		7,645	
John M. Brown	Boston	3,742X	Rev'd John Brown
John B. Brown	Boston	3,474X	
John Mills Brown	Boston	4,370X	
William Gibbs	Prescott	5,927	William Gibbs
William P. Gibbs	Boston	3,133	William Gibbs
Enoch Hale	Boston	7,427X	Enoch Hale
Joseph Howe	Boston	1,840X	Joseph N. Howe, Jr.
J.H. Lord[9]	Boston	65	John Lord
		98	
George W. Lyman	Boston	9,098X[10]	George W. Lyman
William Mason	Taunton	724	William Powell Mason
		1,801	
		4,779	
Daniel Parker	Hubbardston	8,561	Daniel Pinckney Parker

Name on Patent	City	Patent Number	Possible Match
Samuel Parker	Bellerica	900X	Samuel P. Parker
		901X	
		1,038X	
		1,907X	
William W. Parrott	Boston	142	William Parrott
Wm. Perkins	Boston	2,029	William Powell Perkins
George Pratt	Boston	5,905	George Pratt
William Prescott	Boston	*7,082X*	William Prescott
James S. Savage	Boston	607	James Savage
M.H. Simpson	Boston	1,161	Michael Hodge Simpson
Michael H. Simpson	Boston	8,759X	
Amos F. Smith	Salem	*5,642X*	Rev'd Amos Smith
J. Wright Warren, Jr.	Boston	1,710	J.W. Warren, M.D.
I.A. Winslow	Roxbury	7,587	Isaac Winslow

[9]Co-inventor with Enoch Robinson

[10]George W. Lyman was the assignee of patent 9098X, Wool and flax combing machine, issued to William W. Calvert on September 18, 1835.

Appendix F

Transcriptions of Charges

F.1 Charges of Nathaniel Bowditch

Date	Shelf	Charge
1827		
Feb. 27		Mullner. v.1.
		Ibid. v.2.
Mar. 14		*Ibid.* v.3.
		Ibid. v.4.
Mar. 24	1137	Wilson. *Lights & Shadows of Scottish Life.*
Mar. 31		Körner. *Sämmtliche Werke.* v.1, v.2.
		Ibid. v.3.
	1136	Opie. *Simple Tales.* v.1.
May 30	1336	Klopstock. *Werke.* v.8–v.10.
		Wieland. *Oberon, A Poem.* v.23, v.24.
June 11	1139	Austin. *Chironomia, A Treatise on Rhetorical Delivery.*
June 18		Wieland. *Oberon, A Poem.* v.5, v.6.
		Museus. v.1, v.2.
July 6	1346	*Ibid.* v.3, v.4, v.5.

Date	Shelf	Charge
July 23	1139	Austin. *Chironomia, A Treatise on Rhetorical Delivery.*
	1344	La Motte-Fouqué. *Kleine Romane.* v.1, v.2.
Aug. 12	1323	Toqua. *Plays.*
	1323	*Ibid. Zaubining.* v.1.
Aug. 20	1129	Austin. *Chironomia, A Treatise on Rhetorical Delivery.*
	1323	Toqua &c. 12mo, v.2, v.3
Aug. 25	1320	Engel. *Sämmtliche Werke.* v.5.
	1320	*Ibid.* v.12
Aug. 31	1323	Grimm. *German Popular Stories.* v.1
	1323	*Ibid.* v.2
Sept. 18	1108	Butler. *Reminiscences.*
Oct. 4		*Philosophical Transactions.* 1820.
Oct. 6	1139	Austin. *Chironomia, A Treatise on Rhetorical Delivery.*
Oct. 27	1133	Opie. *Temper.* v.1.
Nov. 3	1133	*Ibid.* v.2.
Nov. 17	1136	Butt. *Spanish Daughter.* v.1.
Nov. 24	1139	Austin. *Chironomia, A Treatise on Rhetorical Delivery.*
	1136	Butt. *Spanish Daughter.* v.2.
Dec. 15	1138	Redwood. *A Tale.* v.1.
Dec. 19	1138	*Ibid.* v.2.
Dec. 21		*Sketch Book.* v.1.
Dec. 28		*Ibid.* v.2.
Dec. 29		Irving. *Bracebridge Hall, or the Humorists.* v.1.
1828		
Mar. 29		Sandham. *Twin Sisters, or the Advantages of Religion.*
Apr. 9		Sparks. *Life of John Leulyard.*
Apr. 11		Scott. *Life of Napoleon.*
		Mme. de. Staël Holstein. *Corinne, ou l'Italie.*
July 20	89	Bigelow. *Florula Bostoniensis.*

Date	Shelf	Charge
Aug. 1	1324	Wieland. *Oberon, A Poem.* v.15, v.16.
Aug. 15		*Life of Napolean.*
1829		
Jan. 5	1566	Cooper. *Lectures on the Principles and Practice of Surgery.* v.1.
Jan. 24	1566	*Ibid.* v.2.
Mar. 11	19	Laplace. *Mécanique Céleste.* v.1
June 17	19	*Ibid.*
Oct. 5		Howell. *The life and adventures of Alexander Selkirk.* (Lent to Prof. Farrar).
Nov. 20		Racine. *Oeuvres.*
Dec. 17		Corneille. *Théatre de, avec des Commentaires.* v.1.
Dec. 19	569	Turner. *History of England.* v.1.
Dec. 28	569	*Ibid.* v.2.
1830		
Jan. 27		Voltaire. *Oeuvres Completes.*
Jan. 29		Turner. *History of England.* v.3.
Feb. 13	19	Laplace. *Mécanique Céleste.* v.2.
Feb. 24		Corneille. *Théatre de, avec des Commentaires.* v.2.
Feb. 26	628	Astle. *Origin and Progress of Writing.*
Mar. 6		Tilloch & Taylor. *Philosophical Magazine.* v.25, v.26, v.27.
Mar. 9		*Cambridge Transactions.* v.2.
Mar. 12	1315	Voltaire. *Oeuvres completes.* v.1
Mar. 13	1315	*Ibid.* v.3.
Mar. 18	1315	*Ibid.* v.2.
Mar. 27	1567	Plumbe. *On Ringworm of the Scalp, Scalled Head. and Other Species of Porrigo.*
June 12	19	Laplace. *Mécanique Céleste.* v.2.
Nov. 2		Tilloch & Taylor. *Philosophical Magazine.* v.63, v.67, v.68.
		Ibid. N.S. v.1, v.2, v.3.
		Ibid. June 1830.

Date	Shelf	Charge
1831		
Jan. 26		*Ibid.* 1827.
Feb. 17	1118	Hayley. *The Life and Posthumous Writings of Wm. Cowper.* v.1, v.2.
Feb. 25	1118	*Ibid.* v.3, v.4.
Mar. 4	110	Hill. *The British Herbal: An History of Plants and Trees.*
Mar. 10		Chandler. *Ways of Flesh.*
Mar. 25	626	Simond. *Tour in Italy and Sicily.*
Sept. 2	1524	Lang. Exp.
Sept. 3	682	*Family Library.* v.14
Sept. 20	1546	Foot. *On Curing Diseases of the Urethra Bladder.*
	1546	Bingham. *On Diseases and Injuries of the Bladder.*
Nov. 18		*Correspondence sur École Polytechnique.* v.2
Nov. 21		*Ibid.* v.3
Nov. 24		James. *Nature and Treatment of the Different Species Of Inflammation.*
		Wilson. *Lectures on the Blood.*
Dec. 3		Scott. *Rob Roy.*
1832		
Mar. 2	1296	Fenelon. *Oeuvre.* v.3.
June 9	466	Canning. *Speeches.* v.1.
June 9		*Transactions of the Royal Society of Turin.* v.31, v.32.
June 13	466	Canning. *Speeches.* v.2.
June 20	466	*Ibid.* v.3.
	466	*Ibid. Supplements*
June 25	829	*Englishman's Magazine.* v.1.
July 4	828	*Westminster Review.* v.13.
	534	*History of the French Revolution.* v.1.
July 10	534	*Ibid.* v.2.

Date	Shelf	Charge
July 18	829	*Metropolitan Magazine.* v.1.
July 26	534	*History of the French Revolution.* v.3.
	829	*Metropolitan Magazine.* v.2.
Aug. 6	19	Laplace. *Mécanique Céleste.* v.3.
Aug. 8	799	*Blackwood Magazine.* v.30.
Aug. 23	677	Sparks. *Life of John Leulyard.*
	677	Hunt. *Lord Byron and some of his Contemporaries.*
Sept. 3	677	Irving. *Calumny Supplanted.* v.1, v.2.
Sept. 11	677	*Ibid.* v.3.
Sept. 15	694	Moore. *Sheridan.*
	678	Wollerych. *Jefferys.*
Sept. 15	379	Butler. *The Analogy of Religion.*
Sept. 24	675	Field. *Memoirs of Rev. Samuel Parr, LLD.*
Oct. 11	675	Barrington. *Sketches.*
Oct. 24	183	*Phrenological Journal.* v.3.
	884	Fletcher. *History of Poland.*
Nov. 7	703	Lavalette. *Memoirs of Lavalette.* v.1, v.2.
Nov. 14	884	*Spain in 1830.* v.2.
Nov. 24	783	*Ladies Family Library.* v.1.
	1108	Kelly. *Reminisences.* v.1, v.2.
Dec. 5	673	James. *History of Charlemagne.*
	1118	*Epistles.*
Dec. 19	783	*Ladies Family Library.* v.2.
	544	*England & France.*
1833		
Jan. 7	19	Laplace. *Mécanique Céleste.* v.3.
	379	Butler. *The Analogy of Religion.*
	693	Campbell. *Memoirs.* v.1, v.2.
Jan. 18	863	Stocqueler. *Fifteen Months' Pilgrimage through Untrodden Tracts of Khuzistan and Persia*
Jan. 30	724	Morrill. Voyage.

Date	Shelf	Charge
	863	Dalrymple. *Memoir of his Proceedings as Connected with the Affairs of Spain*
Feb. 14	886	Earle. *New Zealand.*
	case	Byron. *Voyage South Sea.*
Feb. 23	374	Butler. *The Analogy of Religion.*
	case	Beechy. *Voyage.* v.1, v.2.
Feb. 25	618	Cook. *Voyage.* v.1.
Mar. 6	628	Franklin. *Polar Sea, Second Expedition.*
Mar. 15	618	*Cook's Voyages.* v.2.
Mar. 30	863	Monchear. *Letters.* v.1, v.2.
Apr. 6	702	Burney. *Memoirs of Doctor Burney.*
Apr. 11	1148	Campbell. *British Poets.* v.1.
Apr. 20	618	*Cook's Voyages.*
	552	Dumont. *Recollections of Mirabeau.*
Apr. 24		*Transactions of the Berlin Academie* 1824.
Apr. 27		*Mémoires de Turin.* v.35.
May 3	702	Burney. *Memoir of Dr. Burney.* v.3.
May 27	764	Graham. *History of the United States.* v.1, v.2.
June 12	509	Turner. *History of England.* v.1.
June 15	374	Turner. *Sacred History.* v.2.
	708	Las Cases. *Napoleon.* v.1.
June 17	508	Kelly. *The Universal Cambist and Commercial Instructor.* v.1, v.2.
June 29	708	Las Cases. *Napoleon.* v.2, v.3.
July 9	708	*Ibid.* v.4.
July 16	696	O'Meara. *Napoleon.* v.2.
July 26	1263	Nugent. *Legends of the Library at Lilies.* v.1.
Aug. 1	703	*Governor Morris's Life.* v.2.
Aug. 20	764	Duhring. *Remarks on the United States of America.*
	883	*German Prince in England.* v.1.
Aug. 23	684	Bourrienne. *Napoleon.* v.1.

Date	Shelf	Charge
Aug. 26	684	*Ibid.*, v.2, v.3.
Sept. 2	677	Fouché. *Memoirs of Joseph Fouché.*
Sept. 6	684	Bourrienne. Napoleon. v.4
Sept. 9	883	*German Prince in England.* v.2.
Sept. 10	883	*Ibid.* v.4.
Sept. 11	598	*German Prince in England.* (Am. Ed.)
Sept. 13	610	Fisher. *A Journal of a Voyage of Discovery.*
Sept. 21	934	Ségur. *History of the Expedition to Russia.*
	692	Boaden.*Memoirs of Mrs. Inchbald.* v.1.
Oct. 3	692	*Ibid.* v.2.
Oct. 7	782	Hamilton. *Men and Manners in America.* v.1, v.2.
Oct. 14	783	Vigne. *Six Months in America.* v.1, v.2.
Oct. 18	883	Cox. *Adventures on the Columbia River.* v.1, v.2.
Oct. 25	782	*Life of John ???.* v.1.
	682	*Cushing's Reminiscences.* v.1.
Nov. 5		*Life of Fredericke, II.* v.2.
	723	Cushing. *Spain.* v.2.
Nov. 16	682	Pellico. *Memoirs of Silvio Pellico.*
	713	Transatlantic Sketches. v.1
Nov. 23	692	Roscoe. *Life of William Roscoe.* v.1, v.2.
Dec. 3	568	Henderson. *Iceland.* v.1, v.2.
Dec. 6	677	Irving. *Voyages and Discoveries of the Companions of Columbus.*
Dec. 13	702	Finich. *Travels in the United States.*
Dec. 14		Irving. *Voyages and Discoveries of the Companions of Columbus.* v.3.
		Greg. *Sketches of Turkey & Greece.*
		Mrs. Child's Appeal.
Dec. 21	1249	Galt.*Life of Lord Byron.*
	764	Thatcher. *Indian Biography.*
Dec. 25		Taylor. *Life of Cooper.*

Date	Shelf	Charge
1834		
Jan. 14	884	Aikin. *Memoirs of the Court of Queen Elizabeth.* v.1, v.2.
	723	Walpole. *Letters of Horace Walpole.* v.3.
Jan. 23	884	De Kay. *Sketches of Turkey.*
		Memoirs of Baron Cuarier by Mrs. Lea.
Feb. 20	693	Gordon. *Memoirs of Himself.* v.1, v.2.
Mar. 1	1213	Maxwell. Wild Sports of the West. v.1, v.2.
Mar. 1	723	Nugent. *Some Memorials to John Hampton.* v.2.
Mar. 7	682	*Memoir of Roger Williams.*
Mar. 19	783	*Sketches and Eccentricities of Col. David Crockett.*
Mar. 25	1219	*Journal the Hostage.* v.1, v.2.
Mar. 27	1219	*Dutchman's Fussille.* v.1, v.2.
Apr. 2	1214	Hamilton. *Cottagers of Glenburnie.*
Apr. 24	584	*Dr. Beatrice's Residence in Germany.* v.1, v.2.
May 3	462	Flamank. *Treatise on Happiness.* v.1, v.2.
	884	*Italian Exile in England.*
May 31	447	Vaughn. *Memorials of the Stuart Dynasty.* v.1, v.2.
June 21	722	Walpole. *Letters of Horace Walpole.* v.1.
June 31	552	*Abrantès. Memoirs of the Duchess d' Abrantès.*
July 15	18	Laplace. *Mécanique Céleste.* v.4.
	1145	Swift. *Works.* v.10.
Sept. 16		Mawe. *Linnæan System of Conchology.*
Nov. 28	559	*Lady Russell's Letters.*
	1148	*Mrs. Norton's Poems.*
Dec. 25	782	*Ladies Family Library.* v.1, v.2.
1835		
Jan. 7	896	*History of Greece.*
Jan. 16	1234	*Plutarch's Lives.*
	682	Carlyle. *Life of Schiller.*
	18	Laplace. *Mécanique Céleste.* v.4.

Date	Shelf	Charge
Jan. 25	274	Mawe. *Linnæan System of Conchology.*
	896	*History of Greece.*
Feb. 18	1234	*Plutarch's Lives.*
June 4	1320	Taylor. *Plato.*
		Godman. *Natural History.* v.2.
June 29		*Transactions of the Royal Society of Edinburgh.* v.11.
Aug. 14	74	Lyell. *Principles of Geology.*
Oct. 23	1140	Irving. *Conquest of Granada.* v.2.
Oct. 24		Macintosh. *History of England.* v.2.
Nov. 4	74	Lyell. *Principles of Geology.*
Dec. 28	552	Dumont. *Recollections of Mirabeau.*
1836		
Feb. 24		*Biographie Universelle.* v.21.
Feb. 26	562	Turnbull. *French Revolution of 1830.*
Mar. 5	535	Wilberforce. *On the Subject of the Slave Trade.*
	672	*Memoir of Howard.*
Mar. 30	13	Zach. *Correspondence Astronomique, Géographique, Hydrographique.* v.13, v.14.
Apr. 11		Gottingham. *Gothic Ornaments.*
Apr. 13	1715	Pugin. *Specimens of Gothic Architecture.*
	1715	*Ibid. Gothic Examples.*
July 4	1308	Rousseau. *Works.*
Aug. 9	1118	Cowper. *Poems.* v.9.
Aug. 25	1323	Austin. *Characteristics of Goethe.* v.1.
Sept. 5	672	Jamison. *Biography of Women.* v.1, v.2.
Sept. 6	1323	Austin. *Characteristics of Goethe.* v.2.
Sept. 23	472	Smith. *Progress of Philosophy.*
Sept. 28	1320	Taylor. *Plato.* v.2.
Oct. 7		*Cambridge Philosophical Transactions.* v.5.
Nov. 18		Austin. Characteristics of Goethe. v.3.
Dec. 14	125	Cuvie. *Lectures on Comparative Anatomy.* v.1, v.2.
	125	*Ibid.* Atlas.

Date	Shelf	Charge
1837		
Feb. 11	1135	Owenson. *Wild Irish Girl.* v.1, v.2.
	563	Foxe. *Speeches.* v.3.
Feb. 25		Macintosh. *England.* v.2.
Sept. 3	1135	Owenson. *Wild Irish Girl.* v.1, v.2.
Sept. 16		*Edinburgh Review.* v.39.
Oct. 9		*Ibid.* v.48.
Nov. 18	1328	Hayward. *Goethe's Faust.*
Dec. 14	1536	Hepennates. *Air.* v.4.
1838		
Jan. 19		*Edinburgh Review.* 1820. v.43.
Feb. 19	763	James. *Account of an Expedition from Pittsburgh.* v.1, v.2.
Feb. 26	803	Hall. *Schlons Heinfelt*
Mar. 3	1140	Irving. *Conquest of Granada.* v.1, v.2.
Mar. 9	682	Carlyle. *Life of Schiller*
	323	*Life of Sir J. Macintosh* v.1
	595	*Oxford Guide.*
Mar. 13	323	*Life of Sir J. Mackintosh* v.2.

F.2   Periodical Charges of Uriah A. Boyden

Under each journal is the title of an article that may have interested Boyden.

Date	Journal/Article	Volume
1841		
Mar. 19	*Journal of Science* General principles of the Resistance of Fluids, pp. 135–39	1834, v.27
Mar. 19	*Report of BAAS* State of the Analytical Theory of Hydrostatics and Hydrodynamics, pp. 131–52	1833, 3rd Mtg.
Mar. 20	*Philosophical Magazine* Figure of Equilibrium of a Fluid, pp. 321–24	1838, v.13
Apr. 27	*Report of BAAS* Address by the Rev. W. Vernon Harcourt, pp. 3–128	1839, 9th Mtg.
May 8	*Memoirs Philosophical Society of Manchester* Measure of Moving Force, pp. 105–258	1813, v.2 (2s)
May 31	*Annales de Chimie et de Physique* Expériences Relatives à la Résistance que l'Eau Éprouve en se Mouvant dans des Conduites, pp. 244–55	1830, v.43
May 31	*Journal de École Polytechnique* Numerical methods for differential equations by Ampère and Poisson	1820, v.11
June 14	*Report of BAAS* On the Turbine Water-wheel, pp. 191–92	1840, 10th Mtg.

Date	Journal/Article	Volume
June 18	*Thomson's Annals of Philosophy* Œrsted, Critical and Analytical Account of Recherches sur l'Identité des Forces Chimiques et Electriques, pp. 368–77, 456–63	1819, v.13
Sept. 1	*Report of BAAS* See Mar. 19	1833, 3rd Mtg.,
Sept. 3	*Report of BAAS* See June 14	1840, 10th Mtg.
Oct. 7	*Robinson's Mechanical Philosophy*	v.2
Oct. 7	*Robinson's Mechanical Philosophy*	Plates
Dec. 7	*Nicholson's Journal of Philosophy* Description and Account of a New Press operating by the Action of Water on the Principle of the Hydrostatic Paradox, pp. 29–31	1797, v.1
Dec. 11	*Philosophical Magazine* See Mar. 20	1838, v.13
Dec. 24	*Philosophical Transactions London* Use of Tables of Natural and Logarithmic Sines, Tangents, &c in the Numerical Resolution of Affected Equations, pp. 454–78	1781, v.71
Dec. 24	*Philosophical Transactions London*	179(?)
1842		
Aug. 24	*Mémoires de la Société d'Arcueil* Guy-Lussac, Pour déterminer les variations de tempéature qu'éprouvent les gaz en changeant de densité , et considérations sur leur capacité pour le calorique, pp. 180–203	1807, v.1

Date	Journal/Article	Volume
Aug. 24	*Annales de Chimie et de Physique* Guy-Lussac, Sur la Décomposition du Carbonate de Chaux au Moyen de la Chaleur, pp. 219–24	1836, v.63
Aug. 24	*Annales de Chimie et de Physique* Laurent, Recherchés sur les Combustibles minéraux, pp. 337–65	1837, v.64
Sept. 8	*Thomson's Annals of Philosophy* A New Hydraulic Machine, pp. 412–15	1813, v.2
Sept. 24	*Annales de Chimie et de Physique* Dumas, Recherches sur le véritable poids atomique du carbone, pp. 5–59	1841, v.1 ns
Sept. 24	*Annales de Chimie et de Physique* Poiseuille, Recherches expérimentales sur le mouvement des liquides dans les tubes de très-petits diamètres, pp. 74–86	1843, v.85 (v.9 3s)
Oct. 12	*Memoires de l'Academie Royale des Sciences (Turin)*	1829, v.33
Dec. 6	*Annales de Chimie et de Physique* Becquerel, Des lois du dégagement de la chaleur pendent le passage des courants électriques à travers les corps solides et liquides, pp. 21–70	1835, v.81
Dec. 6	*Annales de Chimie et de Physique* Poiseuille, Recherches expérimentales sur le mouvement des liquides dans les tubes de très-petits diamètres, pp. 74–86	1843, v.83 (v.7 3s)

Date	Journal/Article	Volume
Dec. 26	*The Boston Journal* Treadwell, Inventions Connected with Navigable Canals, pp. 473–91, 580–87	1824, v.1
Dec. 26	*Philosophy Magazine* Sylvester, Memoir on Rational Derivation from Equations of Coexistence, that is to say, a new and extended Theory of Elimination, pp. 428–35	1839, v.15
Dec. 26	*Tilloch's Philosophical Magazine* On the Resistance of Cast-Iron in relation to its Use for Conduit Pipes and the Boilers of Steam-Engines, pp. 270–75	v.61
Dec. 26	*Edinburgh Philosophical Journal* Leslie, Enumeration of the Instruments requisite for Meteorological Observations, pp. 141–45	1827, v.2
1843		
Jan. 2	*Philosophy Magazine* Henry, Contributions to Electricity and Magnetism. No. III. On Electro-dynamic Induction, pp. 200–10	1840, v.16
Jan. 2	*Philosophy Magazine* Faraday, On Magneto-electric Induction; in a Letter to Mr. Gay-Lussac, pp. 281–89	1840, v.17
Feb. 20	*Philosophical Transactions London* See Dec. 24, 1841	1781, v.71
Mar. 10	*Mémoires de l'Académie de l'Institut*	v.49

Date	Journal/Article	Volume
Mar. 10	*Nicholson's Journal* De Luc, On the Electric Column and Aerial Electroscope, pp. 81–98, 161–73, 241–68	1810, v.27
Mar. 10	*Nicholson's Journal* De Luc, On Hygrology, Plygrometry, and their connexions with the Phenomena observed in the Atmosphere, pp. 221–303	1812, v.33
Mar. 24	Thomson's *Annals of Philosophy* See June 18, 1841	1813, v.2
Apr. 8	*The Boston Journal* See Dec. 26, 1842	1824, v.1
Apr. 8	Tilloch's *Philosophical Magazine* See Dec. 26, 1842	v.61
Apr. 13	Memoirs Philosophical Society of Manchester See May 8, 1841	1813, v.2 (2s)
July 17	*Transactions Cambridge Philosophical Society* Challis, Researches in the Theory of the Motion of Fluids, pp. 173–204	1835, v.5
Aug. 26	*Report of BAAS* See Mar. 19, 1841	3rd Meeting, 1833
Oct. 28	*Transactions of the Royal Society Edinburgh* Forbes, Experimental Researches regarding certain Vibrations which take place between Metallic Masses having different Temperatures, pp. 429–61	1834, v.12

Date	Journal/Article	Volume
1845		
May 13	*Berlin Academie*	1750, v.6
	Euler, Decouverte d'un nouveau principe de Mécanique, pp. 185–217	
June 23	*Berlin Academie*	1751, v.7
	Euler, Application de la machine hydraulique de M. Segner a toutes sortes d'ouvrages et de ses avantages sur les autres machines hydrauliques dont on se sert ordinairement, pp. 271–304	
July 2	*Annales de Chimie et de Physique*	1819, v.11
	Prony, Des Marais Pontins, pp. 126–79	
1847		
Jan. 2	*Berlin Academie*	v.10
	Euler, Théorie plus complette des machines qui sont mises en mouvement par la reaction de l'eau, pp. 227–295	
Mar. 30	*Franklin Journal*	1830, v.9
	For an improvement in the Application of Hydraulic Power; Zebulon and Austin Parker, Coshocton county, Ohio, October 19, pp. 33–34	
Apr. 12	*Annales de Chimie et de Physique*	1845, v.13 (3s)
	Bravais and Martins, De la Vitesse du Son entre deux stations également ou inégalement élevées au-dessus du niveau de la mer, pp. 5–29	

F.3 Patent Journal Charges of Robert H. Eddy

Date	Journal	Volume(s)
1837		
Feb. 6	*Repertory of Patent Inventions*	10
Feb. 18	*London Journal of Arts & Sciences*	2S 4
Mar. 18	*Repertory of Patent Inventions*	NS 1, 2, 3, 14, 15, 16
Mar. 20	*Repertory of Patent Inventions*	NS 8, 9, 10, 11, 12, 13
Mar. 21	*Repertory of Patent Inventions*	NS 1,2
Mar. 22	*Repertory of Patent Inventions*	2S 42, 43, 44, 45, 46
Apr. 7	*London Journal of Arts & Sciences*	6
Sept. 26	*London Journal of Arts & Sciences*	7,8
1838		
Aug. 15	*London Journal of Arts & Sciences*	2S 4
Aug. 24	*Repertory of Patent Inventions*	8,9
Aug. 25	*Repertory of Patent Inventions*	14,15
1839		
June 25	*London Journal of Arts &d Sciences*	CS 10
June 25	*London Journal of Arts & Sciences*	1S 10
1840		
Feb. 12	*London Journal of Arts & Sciences*	2S 4
Apr. 23	*Repertory of Patent Inventions*	1S 1826/3
Apr. 25	*Repertory of Patent Inventions*	1824/44, 1829/8
Apr. 28	*London Journal of Arts & Sciences*	NS 10
Sept. 15	*London Journal of Arts & Sciences*	5
Oct. 28	*London Journal of Arts & Sciences*	5
Oct. 2	*London Journal of Arts & Sciences*	2, 4, 6, 9, 10, 12
Oct. 15	*London Journal of Arts & Sciences*	7,12

Date	Journal	Volume(s)
1841		
Jan. 19	*London Journal of Arts & Sciences*	CS 9
Feb. 9	*London Journal of Arts & Sciences*	CS 9
Mar. 25	*London Journal of Arts & Sciences*	2
Mar. 29	*London Journal of Arts & Sciences*	CS 1
Mar. 29	*London Journal of Arts & Sciences*	2S 7
Apr. 14	*London Journal of Arts & Sciences*	2S ??
Apr. 14	*London Journal of Arts & Sciences*	CS 12
May 29	*London Journal of Arts & Sciences*	CS 5
Sept. 27	*London Journal of Arts & Sciences*	CS 5
Dec. 15	*Franklin Journal*	3
1842		
Aug. 17	*London Journal of Arts & Sciences*	CS 12
1843		
July 26	*London Journal of Arts & Sciences*	CS ??
Aug. 25	*London Journal of Arts & Sciences*	CS 3
Nov. 2	*London Journal of Arts & Sciences*	CS 17
1845		
Jan. 27	*Repertory of Patent Inventions*	20
Feb. 3	*Repertory of Patent Inventions*	20
Feb. 3	*London Journal of Arts & Sciences*	??
Mar. 29	*London Journal of Arts & Sciences*	22
Oct. 21	*London Journal of Arts & Sciences*	CS 10
1846		
Sept. 5	*London Journal of Arts & Sciences*	1, 5, 9, 16
1848		
Nov. 16	*Repertory of Patent Inventions*	4S 12
Dec. 26	*Repertory of Patent Inventions*	NS 3, 13

F.4 Periodical Articles of Interest to William Lyman

The following table lists the periodical charges of William Lyman. Following the table is a list of articles in the charged journals that may have been of interest to Lyman.

Date	Journal	Volume(s)
1827		
Oct. 1	*Gill's Technical Repository*	11
Oct. 1	*Newton's Journal of the Arts*	13
1829		
Apr. 8	*Trans. R. Soc. London*	1749–50
Apr. 8	*Trans. R. Soc. London*	1751–52
Apr. 8	*Trans. R. Soc. London*	1753–54
June 24	*Register of the Arts*	3, 4, 1NS
Dec. 19	*Register of the Arts*	1, 2, 3
1830		
July 10	*Franklin Journal*	6, 7
1831		
Jan. 26	*London Mechanics Magazine*	12
June 21	*Franklin Journal*	1
Dec. 10	*Franklin Journal*	1831
1832		
Jan. 14	*Register of Arts and Sciences*	5 [NS]
Jan. 23	*Repertory of Patent Inventions*	9, 10
Jan. 25	*London Journal of the Arts*	5
Feb. 1	*Gill's Technical Repository*	4
Feb. 1	*Trans. R. Soc. London*	1806
Feb. 4	*American Journal of Science*	18, 19
Feb. 13	*American Journal of Science*	13
Feb. 16	*American Journal of Science*	14, 15
Feb. 29	*Boston Mechanics Magazine*	1

Date	Journal	Volume(s)
Mar. 8	*American Journal of Science*	11
June 23	*London Mechanics Magazine*	15
July 14	*Gill's Technical Repository*	6
July 14	*Bull. Soc. d'Enc. Arts*	28
July 17	*Bull. Soc. d'Enc. Arts*	27
Sept. 21	*Repertory of Patent Inventions*	7, 9
Oct. 6	*American Journal of Science*	31
Oct. 6	*Bull. Soc. d'Enc. Arts*	24, 25
Nov. 14	*Boston Mechanics Magazine*	16
Dec. 18	*American Journal of Science*	19
1833		
Feb. 16	*American Journal of Science*	22
Mar. 29	*American Journal of Science*	19
1834		
June 18	*Repertory of the Arts*	16
Oct. 9	*Franklin Journal*	13
1835		
Sept. 12	*Franklin Journal*	18
Sept. 12	*Repertory of Patent Inventions*	19
1836		
Jan. 13	*Trans. Geological Soc. Penn.*	8
Feb. 5	*Repertory of Patent Inventions*	19
Feb. 20	*Franklin Journal*	1835
Feb. 24	*London Mechanics Magazine*	21, 22, 23
Feb. 26	*Franklin Journal*	1835

March 1, 1827 *Gill's Technical Repository*, Volume 11

"Experiments made to determine the comparative quantities of Heat evolved in the Combustion of the principal varieties of Wood and Coal used in the United States of North America for Fuel; and also, to determine the comparative quantities of Heat lost by the ordinary Apparatus made use of for their Combustion." By Marcus Bull, Esq.

April 8, 1829 *Trans. R. Soc. London*, 1749–50, 1751–52, 1753–54

One can only conjecture that Lyman was looking for a paper that somebody told him about or that he ran across in his reading although why he'd check out three volumes to look for an article rather than just looking at the Athenæum is a bit of a puzzle. The article probably wasn't "The Case of Nicolas Reeks, Who Was Born with His Feet Turned Inwards, Which Came to Rights after Being Some Time Used to Sit Cross-Legged" or "A Description of the Great Black Wasp, from Pensylvania [*sic*]."A Remark concerning the Sex of Holly

If I were pressed to pick one article, I'd go with "The Best Proportions for Steam-Engine Cylinders, of a Given Content, Consider'd;" By Francis Blake, Esq; F. R. S. which appeared in the 1751–1752 volume pp. 197–201, but it is a little more mathematical than other articles attributed to Lyman's interest.

June 24, 1829 *Register of the Arts*, 3

"White and Sowerby's Patent Air Furnace for Melting Iron." pp. 70 ff.
"Horton's Patent Process in Manufacturing Wrought Iron." pp. 141 ff.
"Luckcock's Patent Process in Manufacturing Iron." pp. 340 ff.

June 24, 1829 *Register of the Arts*, 4

Volume 4 ended the first series and introduced the "New and Improved Series." The name changed from *Register of the Arts and Sciences* to *Register of the Arts, and Journal of Patent Inventions*. The journal must have been doing well because starting on page 407 there is an editorial by Herbert with the headline "Newton's Journal versus the Register of Arts" that rejoins a suit by Newton against the Register for copying.

"Heathorn's Patent Combination of a Lime Kiln with Coke Ovens." pp. 289 ff.

"Wass's Patent Improved Furnace for Rendering the Smelting of Lead Ores and Other Mineral Substances Innoxious to the Surrounding Neighbourhood." pp. 205 ff.

"Paterson's Blowing Machine." pp. 293 ff.

June 24, 1829 *Register of the Arts*, 1 NS
"Patent Process in Making Iron" by Philip Taylor, Esq. p. 164.

Furnace for Smelting Iron, by Means of Anthracite, by Joshua Malin, Engineer. pp. 293 ff. A reprint from the *Franklin Journal* with commentary.

December 19, 1829 *Register of the Arts*, 2
"Mr. George Chapman's Plan for Consuming the Smoke of Steam-Boilers." pp. 226 ff. Uses the smoke from the steam boiler to heat the air that is blown into an iron furnace.

December 19, 1829 *Register of the Arts*, 3
See June 24, 1829.

July 10, 1830 *Franklin Journal*, Volume 6
Only very rarely does a registry entry note in which series the volume number of a journal references. For our purposes this is not usually a problem because date of the charge can be used to disambiguate the series. Here, however, there are two Volume 6s; one in the original series covering July to December of 1828 and one in the new series covering July to December of 1830. In the 1828 Volume 6 we find "On the blowing of Air into Furnaces by a Fall of Water by the late celebrated WILLIAM LEWIS, M.D." In Volume 6 of the New Series, we find an extended description of the patent granted to Josias Lambert *for an improvement in making iron.* Lambert's improvement had to do with adding salt, potash, and lime to the iron.

July 10, 1830 *Franklin Journal*, Volume 7
Specification of a patent for an improvement in the art of manufacturing Malleable Iron from Pig Metal. Granted to Thomas Cotton Lewis, Pine Creek, Alleghany[*sic*] county, Pennsylvania.

January 26, 1831 *The Mechanics Magazine*, Volume 12
"Comparative Effects of Blowing Hot and Cold Air into Furnaces." This is a short, one-page note by the editor who is of the opinion that "the expense of making the air hot would be found exactly to balance the subsequent gain." He goes on to describe some experiments reported in the Glasgow Chronicle which found that heating the air was quite economically advantageous.

June 21, 1831 *Franklin Journal*, Volume 1
As with Volume 6 above, there are two candidates for Volume 1, the very first volume of 1828 and the first in the New Series of 1831. In Volume 1 of the New Series we find "On the blowing of Air into Furnaces by a Fall of Water by the late celebrated WILLIAM LEWIS, M.D." except by now Mr. Lewis is late but

not celebrated. In the 1826 Volume 1 two premiums that Lyman pursued are described:

> "14. To the inventor of the best constructed Furnace, for consuming anthracite in generating steam, to be applied to steam engines.—A Silver Medal. Certificates will be required of the furnace having been some time in use, of the quantity of coal consumed, and of the effect produced."
>
> "15. To the person who shall have manufactured in Pennsylvania, the greatest quantity of Iron from the ore, using no other fuel but anthracite, during the year ending September 1, 1826. The quantity not to be less than twenty tons.—A. Gold Medal."

December 10, 1831 *Franklin Journal*, 1831
"Proposed plan for Smelting Iron Ore with Anthracite Coal."

January 14, 1832 *Register of Arts and Sciences*, Volume 5 [NS]
"Disputed Claim of Patent-Right, Between Messrs. Cochrane and Galloway, Plaintiffs, and Messrs. Braithwaite and Erricson, Defendants, respecting a Steam Engine Boiler" pp. 72 ff.

Three years after Cochrane & Galloway got a patent for blowing a furnace, Braithwaite & Ericcson got one on the same topic. Cochrane & Galloway sued Braithwaite & Ericcson, claiming the latter stole their idea and thus infringed on their patent. The editor runs double columns of the testimony of the experts, the left-hand column the experts supporting Galloway and the right-hand column the experts supporting Braithwaite. George Brikbeck, the author of one of our steam engine books, was on the left and John Farey, the author of another of our steam engine books, was on the right. The editor notes prior art for both patents and concludes by writing "No cash account in a merchant's ledger could be more accurately balanced than are the professional opinions in this case."

January 23, 1832 *Repertory of Patent Inventions*, Volume 9
Volume 9 of original series is of 1798. Volume 9 of the Second Series is 1806. In the original series Volume 9 we find "Specification of the Patent granted to Mr. William Finch, of Woombourne, in the County of Stafford, Iron-master; for certain new Methods of making Nails and Spikes, of Iron, Copper, and other Metals, by means of certain Machinery put in Motion by the Force of Animals, Water, Winds or Steam, instead of making them by Hand." pp. 390 ff.

January 23, 1832 *Repertory of Patent Inventions*, Volume 10
A long three-part article, Specification of the Patent granted to Samuel Bentham, of Queen-Square-Place, Westminster, in the County of Middlesex, Esquire; for

his invention of various new and improved Methods and Means of working Wood, Metal, and other Materials. pp. 221. ff., pp. 293 ff., and pp. 367 ff.

February 1, 1832 *Gill's Technical Repository*, Volume 4
Specification of a Patent for the Construction of a Furnace for Generating Steam by Anthracite Coal, and for the Use of various Manufactures requiring Intense Heat. Granted to Benjamin B. Howell, Philadelphia, October 14, 1828.

Specification of a Patent, for an Improvement in the Manufacture of Malleable Iron; and of an Improved Bloomery Furnace. Issued to Benjamin B. Howell, Philadelphia, November 6, 1828.

February 1, 1832 *Trans. R. Soc. London*, 1806
"Description of the Mineral Bason in the Counties of Monmouth, Glamorgan, Brecon, Carmarthen, and Pembroke." by Edward Martin.

February 4, 1832 *American Journal of Science*, Volume 18
"On the use of Anthracite in Blacksmiths' Shops; by G. Jones," Tutor in Yale College.

February 4, 1832 *American Journal of Science*, Volume 19
The lead article in this volume is " Notes on a journey from New Haven, Conn., to Mauch Chunk and other Anthracite regions of Pennsylvania" written by the journal's editor, Benjamin Silliman. The twenty-one page article is more of a travelogue than a scientific treatise, Lyman couldn't help but be keenly interested in what a leading scientist of the day had to say about the area of the country in which he was about to invest.

February 13, 1832 *American Journal of Science*, Volume 13
"Chemical analysis and description of the Coal lately discovered near Tioga River, in the State of Pennsylvania;" by William Meade, M. D.

February 13, 1832 *American Journal of Science*, Volumes 14, 15
"On the Geology and Mineralogy of the country near West Chester, Penn.;" by J. Finch, M. C. C. &c. is one possibility in Volume 14 but more likely, given that Lyman checked out Volumes 14 and 15 together, is a long, multi-part article by two of our other readers, Francis Alger and Charles T. Jackson, "A Description of the Mineralogy and Geology of a part of Nova Scotia."

February 29, 1832 *Mechanicks Magazine*, Volume 11
There is a two-page letter to the Editor signed simply Middlesex which describes internal communication in Pennsylvania.

Mauch Chunk Rail-Way, in 1827; with its branches is thirteen and a half miles long; a single track cost $3,050 per mile; connects the Lehigh canal with the coal mines of this dame.

There is also a reprint of a segment of an article in Silliman's Journal about anthracite coal. It isn't mentioned which volume of Silliman. The excerpted article was *A description of the advantages derived from the use of an intermixture of the Anthracite Coal of this country, with the materials made use of in the making of Brick, rendering the burning of them more perfect and uniform, and greatly improving their texture and durability, as practiced last season, at one of the Brick Yards on the North River*, which appeared in Volume 18, June–July, 1830.

March 8, 1832 *American Journal of Science*, Volume 11
"Anthracite Coal of Rhode-Island—remarks upon its properties and economical uses: with an additional notice of the anthracites of Pennsylvania" by the Editor. This is a continuation of "Anthracite Coal of Pennsylvania. Remarks upon its Properties and economical Uses," which appeared in Volume 10.

June 23, 1832 *The Mechanics Magazine*, Volume 15
"Action for Infringement of Patent. Galloway v. Braithwaite and Ericsson." The patent was for a method of blowing a furnace using a steam engine, a topic of keen interest to Lyman. Mr. John Farey, Civil Engineer, was called to testify as an expert witness.

July 16, 1832 *Gill's Technical Repository*, Volume 6
On Improvements in Cutting Screws and Screw Nuts. By Mr. James Jones, Engineer.

October 6, 1832 *American Journal of Science* , Volume 31
A running commentary regarding an item in the Transactions of the Royal Society of Edinburgh by Thomas Clark, " On the Application of the Hot Blast in the Manufacture of Cast-iron."

November 14, 1832 *The Mechanics Magazine*, Volume 16
State of the Collieries—Necessity of a System of Ventilation.

December 18, 1832 *American Journal of Science*, Volume 19
Lyman revisits Volume 19. He may be refreshing his memory about the article about Mauch Chunk but there is another possibility: "Observations on the Coal Formations in the State of New York; in connexion [*sic*] with the great Coal Beds of Pennsylvania. From the Transactions of the Albany Institute;" by Amos Eaton, Corresponding Member.

February 16, 1833 *American Journal of Science*, Volume 22

A very long article: "Report[1] on the Geology of Massachusetts; examined under the direction of the Government of that State, during the years 1830 and 1831;" by Edward Hitchcock, Prof. of Chemistry and Natural History In Amherst College. The article includes a particularly fine map of the geology of Massachusetts.

March 29, 1833 *American Journal of Science*, Volume 19

See December 18, 1832 above.

October 9, 1834 *Franklin Journal*, Volume 13

There's not much doubt about this one: "American Patents for December, with Remarks. For Making Iron and Steel, by the use of anthracite coal; F. W. Geissenhainer, city of New York, December 19." But this is not the only possibility. Lyman was a partner in a puddling furnace on Mill Dam so another patent might have also been of interest: "[I]mprovements in the Puddling and Heating of Iron in the process of manufacturing it; William Jones, Haverstraw, Rockland county, New York; an alien, who has resided two years in the United States, December 16."

September 12, 1835 *Franklin Journal*, Volume 18

Lyman revisits Volume 18. He could be rereading "On the application of the Hot Blast in the manufacture of Cast-iron." By Thomas Clarke, M. D. Professor of Chemistry in Marischall College, Aberdeen. This volume contains Part II of the monumental "Report of the Committee of the Franklin Institute of the State of Pennsylvania for the promotion of the Mechanic Arts, on the Explosions of Steam Boilers" but I haven't detected any interest in exploding steam boilers on Lyman's part.

February 20, 1836 *Franklin Journal*, 1835

There were two volumes of the *Franklin Journal* in 1835, Volumes 15 and 16, both in the New Series. Since the registry just records 1835, we will cover them both.

In Volume 15, there are two articles, one short and one very long, pinning down the issue of using hot air to blow iron furnaces. The short one is "Extract of a Report upon the employment of Hot Air in smelting Iron with Charcoal." By M. E. Guetmard, Engineer in Chief of Mines. And the long, multi-part article is: "To the Board of Directors of Bridges, Public Roads, and Mines, upon the

---

[1] Footnote in journal: Published in this Journal by consent of the government of Massachusetts, and intended to appear also in a separate form, are to be distributed among the members of the Legislature of the same State, about the time of its appearance in this work. It is, we believe, the first example in this country, of the geological survey of an entire State.

Use of Heated Air in the Iron Works of Scotland and England. By M. Dufrenoy, Engineer of Mines. Paris, 1834. (Translated for this Journal, by S.V. Merrick.)"

Volume 15 also includes "On the Use of Anthracite in the Smelting Furnace of Vizille." By M. Robin, Director of the Neiderbrunn Works.

As a side note, Volume 15 notes two patents issued to residents of Pottsville, Pennsylvania: *Smelting iron ore* to Thomas J. Ridgway and "Iron furnaces, anthracite." M. Brooke Buckley. One would like to conjecture Ridgway and Buckley were employees at Lyman's Pioneer Furnace.

Volume 16 includes an abstract from the Ridgway patent that was only noted in Volume 15: Abstract of a Specification of a Patent for a Furnace for Preparing and Smelting Iron Ore, with Anthracite Coal. Granted to Thomas S. Ridgway, Pottsville, Schuylkill county, Pennsylvania, December 17th, 1834. This is followed thirty-four pages later by a "Report on Mr. Thomas S. Ridgway's Smelting Furnace" by the Franklin Institute's Committee on Science and the Arts. Why the Franklin Institute went out of its way to rubbish Ridgway is a puzzle as the editor of the *Franklin Journal* had been the Superintendent of the Patent Office and would soon return as a Patent Examiner. By this time Lyman along with Robert Forbes Bennett had opened their nail factory in Farrandsville so the following might also have been of interest: "Abstract of the Specification of a Patent for a mode or machine for making Wrought Nails, Tacks, or Spikes." Granted to William C. Grimes, York, York county, Pennsylvania, December 17th, 1834.

February 24, 1836 *The Mechanics Magazine*, Volume 21, 22, 23
"A Compound Reciprocating Rotary Steam-Engine, and an Improved Boiler, invented by Simon Fairman, of Lansingburgh, New York. (From the New York *Mechanics' Magazine*)"

February 26, 1836 *Franklin Journal*, 1835
See February 20, 1836 above.

References

Entries for books in the Boston Athenæum's current collection include callnumber and Athena bibId within square brackets and separated by an at sign (@).

[1] Jacob Abbott. *The Rollo Philosophy: Part III. Fire.* Otis, Broaders, Boston, 1st ed., 1843. [PS1000.A8 R77 1843 @ 378243].

[2] ——. *The Rollo Philosophy. Part II. Air.* Gould, Kendall, and Lincoln, Boston, 1845.

[3] Francis Alger. Notes on the Mineralogy of Nova Scotia. *The American Journal of Science and the Arts (Silliman's)*, 12:227–32, June 1827.

[4] Francis Alger and Charles T. Jackson. A Description of the Mineralogy and Geology of a Part of Nova Scotia. *The American Journal of Science and the Arts (Silliman's)*, 14, 15, 16:305–30, 132–60, 201–17, 1828.

[5] ——. Remarks on the Mineralogy and Geology of Nova Scotia. *Memoirs of the American Academy of Arts and Sciences*, 1:217–350, 1832.

[6] ——. *Remarks on the Mineralogy and Geology of the Peninsula of Nova Scotia: Accompanied by a Colored Map, Illustrative of the Structure of the Country, and by Several Views of its Scenery.* E.W. Metcalf and Co, Cambridge, MA, 1st ed., 1832. [4 557.16 J12 @ 273315].

[7] Neil Arnott. *On Warming and Ventilating; With Directions for Making and Using the Thermometer-Stove, or Self-Regulating Fire, and Other New Apparatus.* Longman, Orme, Brown, Green, and Longmans, London, 1 ed., 1838. [PF .Ar6 @ 231497].

[8] Charles Babbage. Letter to Sir Humphry Davy, Bart. P.R.S., on the Application of Machinery to Calculate and Print Mathematical Tables. *The Edinburgh Review*, 120:263–327, July 1834.

[9] Loammi Baldwin, Jr. *Report on the Subject of Introducing Pure Water into the City of Boston.* John H. Eastburn, Boston, 1834. [Tract B1807 @ 500210].

[10] John Baron. *The Life of Edward Jenner, M.D. LL.D., F.R.S. Physician Extraordinary to His Majesty Geo. IV. Foreign Associate of the National Institute of France. With Illustrations of his Doctrines, and Selections from his Correspondence.* Henry Colburn, London, 1st ed., 1838. [5E .J436 .b @ 114959].

[11] Greville Bathe and Dorothy Bathe. *Oliver Evans: A Chronicle of Early American Engineering.* Historical Society of Pennsylvania, Philadelphia, 1935.

[12] Margaret W. Batschelet. *Arithmetick Vulgar and Decimal: with the Application Thereof to a Variety of Cases in Trade and Commerce.* The Scarecrow Press, Metuchen, NJ, 1990.

[13] Jacob Bigelow. *American Medical Botany: Being a Collection of the Native Medicinal Plants of the United States, Containing their Botanical History and Chemical Analysis, and Properties and Uses in Medicine, Diet and the Arts, With Coloured Engravings.* Cummings and Hilliard, Boston, 1st ed., 1817. [$JX +B48 @ 368540].

[14] ———. *A Discourse on Self-limited Diseases. Delivered before the Massachusetts Medical Society, at their Annual Meeting, May 27, 1835.* N. Hale, Boston, 1st ed., 1835. [R708 .B593 @ 410324].

[15] ———. *An Introductory Lecture on the Treatment of Disease. Delivered before the Medical Class at the Massachusetts Medical College in Boston, November 3, 1852.* Ticknor, Reed, and Fields, Boston, 1st ed., 1853. [Adams D239 @ 410329].

[16] David G. Blair. *An Easy Grammar of Natural and Experimental Philosophy: For the Use of Schools; with Ten Engravings.* Solomon W. Conrad, Philadelphia, 4th ed., 1818. [HA .B57 @ 219760].

[17] Miles Bland. *Algebraical Problems: Producing Simple and Quadratic Equations with their Solutions: Designed as an Introduction to the Higher Branches of Analytics.* Printed by J. Smith, Cambridge, UK, 3rd ed., 1820. [H4 .3B61 @ 225378].

[18] Henry Ames Blood. *The History of Temple, N.H.* Geo. C. Rand & Avery, Boston, 1st ed., 1860. [962T24 .B @ 167063].

[19] Charles Knowles Bolton. Social Libraries in Boston. *Publications of the Colonial Society of Massachusetts: Transactions 1908–1909*, 12:332–38, April 1909.

[20] Charles Knowles Bolton and Wendell Barrett. *The Athenaeum Centenary. The Influence and History of the Boston Athenaeum from 1807 to 1907, with a Record of its Officers and Benefactors and a Complete List of Proprietors.* Boston Athenæum, Boston, 1907. [Z733 .B74 @ 49742].

[21] Jean Charles Borda. Mémoire sur l'Écoulement des Fluides par les Orifices des Vases. *Mémoires Académie des Sciences*, pp. 579–607, 1769.

[22] Boston Athenæum Cataloging Department. *Catalogue of the Books in the Boston Athenaeum.* Boston Athenæum, Boston, 1810. [B.A.12 .1 (1810) @ 417841].

[23] ———. *Catalogue of Books in the Boston Athenaeum: To Which are Added the By-Laws of the Institution, and a List of its Proprietors and Subscribers.* Printed by W. L. Lewis, Boston, 1827. [Z881 .B739 1827 @ 302739].

[24] ———. *Catalogue of Books Added to the Boston Athenaeum since the Publication of the Catalogue in January 1827.* Boston Athenæum, Boston, 1829. [Z881.B739 1827 Suppl. @ 402224].

[25] ———. *Catalogue of Tracts, Scientific and Alphabetical Index.* Boston Athenæum, Boston, 1831. [B.A.12 .9 (1831) @ 418248].

[26] ———. *Catalogue of Books Added to the Boston Athenaeum in 1830–1833.* Printed by Eastburn's Press for the Boston Athenæum, Boston, 1834. [Z881 .B739 1830-1833 @ 402226].

[27] ———. *Catalogue of Books Added to the Boston Athenaeum, since the Publication of the Catalogue in January, 1827.* Printed by Eastburn's Press for the Boston Athenæum, Boston, 1840. [Z881 .B739 1840 @ 302740].

[28] ———. *Shelflists, 1849.* Boston Athenæum, Boston, 1849. [B.A. 12 .20 (1849) @ 427051].

[29] Nathaniel Bowditch. *The New American Practical Navigator: Being an Epitome of Navigation.* Edmund M. Blunt, Newburyport, MA, 1st ed., 1802. [VK555 .B68 1802a @ 9101]. Reference: John F. Campbell. *History and Bibliography of The New American Practical Navigator and the American Coast Pilot.* Salem, MA: Peabody Museum, 1964. p. 76.

[30] ———. *The New American Practical Navigator: Being an Epitome of Navigation.* Edmund M. Blunt for Jacob Richardson, Newburyport, Mass, 1st ed., 1802. [VK555 .B68 1802 @ 9098]. Not mentioned in John F. Campbell. *History and Bibliography of The New American Practical Navigator and the American Coast Pilot.* Salem, MA: Peabody Museum, 1964.

[31] ——. Works of the German Astronomers. *The North-American Review and Miscellaneous Journal*, 1(2):260–72, April 1820.

[32] ——. *Proposals of the Massachusetts Hospital Life Insurance Company: To Make Insurance on Lives, to Grant Annuities on Lives and in Trust, and Endowments for Children*. Printed by James Loring, Boston, 1st ed., 1823. [D57 no.2 @ 340803].

[33] Uriah A. Boyden. Comment on Mr. Lester's Vindication of his Pendulum Steam Engine. *Mechanics' Magazine and Journal of Public Internal Improvement*, 1(10):294–96, November 1830. Signed "U.A.B.".

[34] ——. Letter to the Editor in re "Combustion of Ashes". *Mechanics' Magazine and Journal of Public Internal Improvement*, 1(9):271–72, October 1830. Signed "U.A.B.".

[35] ——. On the Effects of Blowing Hot and Cold Air into Furnaces. *Mechanics' Magazine and Journal of Public Internal Improvement*, 1(7):217–19, August 1830. Signed "U.A.B.".

[36] ——. Vitality of Matter. *Mechanics' Magazine and Journal of Public Internal Improvement*, 1(12):369–71, January 1831. Signed "U.A.B.".

[37] ——. Hanging Shaft of Water-Wheels. US Patent No. 5,068, 1847.

[38] Robert Boyle. Of its Usefulness to the Minde of Man. In *Some Considerations Touching the Usefulnesse of Experimental Natural Philosophy, Propos'd in a Familiar Discourse to a Friend, by way of Invitation to the Study of it*, pp. 1–20. Printed by Henry Hall, Printer to the University for Ric. Davis, Oxford, UK, 2nd ed., 1664.

[39] ——. *Experimentorum novorum physico-mechanicorum continuatio secunda. In qua experimenta varia tum in aere compresso, tum in factitio, instituta, circa ignem, animalia, &c. Unà cum descriptione machinarum continentur*. Excudebat Milo Flesher, pro Richardo Davis, Londini, 1 ed., 1680. [HA .B69 @ 219786].

[40] ——. A Proëmial Essay, Wherein, with Some Considerations Touching Experimental Essays in General, is Interwoven such an Introduction to all those Written by the Author, as is Necessary to be Perused for the Better Understanding of Them. In *The Works of the Honorable Robert Boyle in Six Volumes*, vol. 1, pp. 299–318. J. and F. Rivington, London, 1772.

[41] Francis B. C. Bradlee. *The Boston and Lowell Railroad, the Nashua and Lowell Railroad, and the Salem and Lowell Railroad*. The Essex Institute, Salem, MA, 1918. [PY5 .B65 b @ 233668].

[42] Nathaniel Jeremiah Bradlee. *History of the Introduction of Pure Water into the City of Boston, with a Description of its Cochituate Water Works.* Alfred Mudge & Son, Boston, 1st ed., 1868. [PK964B6 //B728 @ 234326].

[43] Arthur Wellington Brayley. *A Complete History of the Boston Fire Department, including the Fire-Alarm Service and the Protective Department, from 1630 to 1888.* John P. Dale & Co., Boston, 1st ed., 1889. [RYZ64B .B73 @ 243471].

[44] ———. *Schools and Schoolboys of Old Boston: An Historical Chronicle of the Public Schools of Boston from 1636 to 1844, to which is Added a Series of Biographical Sketches, with Portraits of Some of the Old Schoolboys of Boston.* Louis P. Hager, Boston, 1st ed., 1894. [DQ64B6 /B7 @ 228740].

[45] Barnabé Brisson. *Recueil de 245 Dessins ou Feuilles de Textes Relatifs à l'Art de l'Ingénieur, Extraits de la Première Collection Terminée en 1820, et Lithographiés à l'École Royale des Ponts et Chaussées.* Ecole Royale des Ponts et Chaussées, Paris, 1826. [P/P21 @ 234541].

[46] Chandos Michael Brown. *Benjamin Silliman: A Life in the Young Republic.* Princeton University Press, Princeton, NJ, 1989.

[47] Bill Bryson. *A Short History of Nearly Everything.* Broadway Books, New York, NY, 1st ed., 2003. [Q162 .B88 2003 @ 402806].

[48] Carol Bundy. *The Nature of Sacrifice: A Biography of Charles Russell Lowell, Jr., 1835–64.* Farrar, Straus and Giroux, New York, NY, 1st ed., 2005. [CT275.L6837 B86 2005 @ 418153]. See also B.A.15 .20.142 (2005), sound recording of lecture held at the Boston Athenæum, March 22, 2005.

[49] Edmund Burke. *List of Patents for Inventions and Designs Issued by the United States, from 1790 to 1847, with the Patent Laws and Notes of Decisions of the Courts of the United States for the Same Period.* J. & G.S. Gideon, Washington, DC, 1847.

[50] Florian Cajori. *The Teaching and History of Mathematics in the United States.* Government Printing Office, Washington, DC, 1890. Bureau of Education Circular of Information No. 3, Whole Number 167.

[51] François Callet. *Tables Portatives de Logarithmes, Contenant les Logarithmes des Nombres, Depuis 1 Jusqu'à 10800; les Logarithmes des Sinus et Tangentes.* F. Didot, Paris, stéréotype ed., 1812. [H4L .6C13 @ 218261].

[52] Danilo Capecchi. Over and Undershot Waterwheels in the 18th Century. Science-Technology Controversy. *Advances in Historical Studies*, 2(3):131–39, 2013.

[53] Thomas G. Cary. *Memoir of Thomas Handasyd Perkins; Containing Extracts from his Diaries and Letters. With an Appendix.* Little, Brown, Boston, 1st ed., 1856. [F69 .P46 1856 CT275.P472 C37 @ 11182].

[54] Thomas Chambers, G.N. Eckert, and Samuel L. Reeves. Iron and Coal Statistics: being Extracts from the Report of a Committee to the Iron and Coal Association of the State of Pennsylvania, 1846. *The Franklin Journal*, 42:124–141, July 1846. Volume 12 Third Series.

[55] Robert G. Clason. Some Historical Whats, Hows, and Whys in Teaching Arithmetic. *The Arithmetic Teacher*, 17(6):461–72, 1970.

[56] I. Bernard Cohen. Some Reflections on the State of Science in America during the Nineteenth Century. *Proceedings of the National Academy of Sciences*, 45(5):666–77, 1959.

[57] Warren Colburn. *Intellectual Arithmetic, upon the Inductive Method of Instruction.* James L. Cutler and Co, Bellow-Falls, VT, 1820.

[58] ———. *First Lessons in Arithmetic: An Arithmetic on the Plan of Pestalozzi: With Some Improvements.* Cummings and Hilliard, Boston, 1821.

[59] ———. *Arithmetic: Being a Sequel to First Lessons in Arithmetic.* Cummings and Hilliard, Boston, 1822.

[60] ———. *First Lessons in Arithmetic on the Plan of Pestalozzi.* Cummings and Hilliard, Boston, 1822.

[61] ———. *A Key Containing Answers to the Examples in the Sequel to First Lessons in Arithmetic.* Hilliard & Metcalf, Cambridge, MA, 1823.

[62] ———. *An Introduction to Algebra upon the Inductive Method of Instruction.* Hilliard, Gray, Little, and Wilkins, Boston, 1st ed., 1831. [H4 .C67 @ 225407].

[63] Commissioner of Patents. *A List of Patents Granted by the United States, for the Encouragement of Arts and Sciences, Alphabetically Arranged, from 1790 to 1828...containing the Names of Patentee, their Places of Residence, and the Dates of their Patents.* Printed by S. Alfred Elliot, Washington, DC, 1828.

[64] ———. *A List of Patents Granted by the United States from April 10, 1790, to December 31, 1836, with an Appendix Containing reports on the Condition of the Patent-office in 1823, 1830, and 1831.* Commissioner of Patents, Washington, DC, 1872.

[65] Commonwealth of Massachusetts. Laws of the Commonwealth of Massachusetts passed at the Session of the General Court, began and held at Boston on Wednesday the Seventh Day of January, Anno Domini One Thousand Eight Hundred and Seven, 1807.

[66] Wiliam Cook. A Proposal for Warming Rooms by the Steam of Boiling Water Conveyed in Pipes along the Walls: And a Method of Preventing Ships from Leaking Whose Bottoms are Eaten by the Worms. *Philosophical Transactions of the Royal Society of London*, 43:370–72, 1744.

[67] Josiah Parsons Cooke, Jr. The Numerical Relation between the Atomic Weights, with Some Thoughts on the Classification of the Chemical Elements. *The American Journal of Science and the Arts (Silliman's)*, 17(51):387–407, May 1854.

[68] T. G. Cumming. *Illustrations of the Origin and Progress of Rail and Tram Roads, and Steam Carriages, or Loco-Motive Engines: Also, Interesting Descriptive Particulars of the Formation, Construction, Extent, and Mode of Working Some of the Principal Rail-Ways*. Printed for the author, Denbigh, 1824. [PV .C91 @ 233812].

[69] Charles Cutter. *Catalog of the Library of the Boston Athenæum. 1807–1871*. G.K. Hall, Boston, 1874. [:XN5 +B6562 +2 @ 304477].

[70] John Dalton. *A New System of Chemical Philosophy*. Printed by S. Russell for R. Bickerstaff, Manchester, London, 1st ed., 1808. [HK .D17 @ 216313].

[71] Peter Dear. *Discipline & Experience: The Mathematical Way in the Scientific Revolution*. University of Chicago Press, Chicago, IL, 1995.

[72] William Prescott Dexter. *Tabulæ Atomicæ: The Chemical Tables for the Calculation of Quantitative Analyses of H. Rose: Recalculated for the More Recent Determinations of Atomic Weights, and with other Alterations and Additions*. Little and Brown, Boston, 1850.

[73] Kenneth Dobyns. *The Patent Office Pony: A History of the Early Patent Office*. Docent Press, Boston, 2016.

[74] J.A. Drake. *The Practical Mechanic Comprising a Clear Exposition of the Principles and Practice of Mechanism, with their Application to the Industrial Arts*. J.W. Lukenbach, Philadelphia, 1879.

[75] Pierre Armand Dufrénoy. *On the Use of Hot Air in the Iron Works of England and Scotland. Translated from a Report, Made to the Director General of Mines in France by M. Dufrenoy in 1834*. J. Murray, London, 1st ed., 1836. [QC .D87 @ 235904].

[76] Caleb Eddy. *Historical Sketch of the Middlesex Canal: With Remarks for the Consideration of the Proprietors.* Printed by Samuel N. Dickinson, Boston, 1st ed., 1843. [PQ64M .Ed @ 233307].

[77] Robert H. Eddy. *Report on the Introduction of Soft Water into the City of Boston.* Printed by J. H. Eastburn, Boston, 1836.

[78] Robert H. Eddy and Francis Greenleaf Pratt. *The Eddy Family: Reunion at Providence to Celebrate the Two Hundred And Fiftieth Anniversary of the Landing of John and Samuel Eddy at Plymouth, Oct. 29, 1630.* Cushing, Boston, 1st ed., 1881. [65 .9YEd23 @ 132662].

[79] Editor. Review of the Cambridge Course in Mathematics. *The American Journal of Science and the Arts (Silliman's),* 5:304–326, 1822.

[80] Editors. Intelligence and Remarks. *The North-American Review and Miscellaneous Journal,* 5(15):430–439, September 1817.

[81] ——. Report of the Committee on Useful Inventions. *Massachusetts Agricultural Repository and Journal,* 10(1), July 1827. Part of the Official Reports of the Committees of the Massachusetts Agricultural Soceity, as Announced at their Aniversary held at Brighton, October 20, 1830.

[82] ——. Lester's Pendulum Engine. *Mechanics' Magazine and Journal of Public Internal Improvement,* 1(9,10):273–77, 289–96, 1830.

[83] ——. Lester's Pendulum Steam-Engine. *The Mechanics' Magazine, Museum, Register, Journal, and Gazette,* 15(398):50–53, 1831.

[84] ——. *The Boston Mechanic, and Journal of the Useful Arts and Sciences,* vol. 4. Light & Stearns, Boston, 1835.

[85] ——. Review of *An Elementary Treatise on Mineralogy; Comprising an Introduction to the Science. Fifth Edition from the Fourth London Edition,* by Robert Allen; containing the Latest Discoveries in American and Foreign Mineralogy; with Numerous Additions to the Introduction. *The North-American Review and Miscellaneous Journal,* 59(124):240–43, July 1844.

[86] ——. The Ether Controversy. *Littell's Living Age,* 17:491–521, April, May, June 1849.

[87] ——. Review of *First Lessons in Language* by David Tower. *The North-American Review and Miscellaneous Journal,* 79(164):259–60, 1854.

[88] ——. Philosophy in Court. *The Massachusetts Teacher and Journal of Home and School Education*, 10:391–92, 1857.

[89] ——. Review of *Intellectual Arithmetic, upon the Inductive Method of Instruction*. *Popular Science*, 25:850, October 1884.

[90] ——. *Oxford English Dictionary*. Clarendon Press, Oxford, UK, 2nd ed., 2004.

[91] Theodore Edson. *Memoir of Warren Colburn, Written for the American Journal of Education*. Brown, Taggard & Chase, Boston, 1st ed., 1856. [65 .C673 @ 178874].

[92] ——. Warren Colburn. *The American Journal of Education*, 2:294–316, 1856.

[93] Clark A. Elliott. *Thaddeus William Harris 1795–1856: Nature, Science, and Society in the Life of an American Naturalist*. Lehigh University Press, Bethlehem, 1st ed., 2008. [CT275.H3736 E44 2008 @ 449382].

[94] George B. Emerson. *The Schoolmaster. The Proper Character, Studies, and Duties of the Teacher with the Best Methods for the Government and Instruction of Common Schools, and the Principles on which Schoolhouses Should be Built, Arranged, Warmed and Ventilated*, pp. 265–538. Harper and Brothers, New York, NY, 2nd ed., 1842.

[95] Leonhard Euler. Decouverte d'un Nouveau Principe de Mécanique. *Mémoires de l'Académie des Sciences de Berlin*, 6:185–217, 1752. Presented to the Berlin Academy on September 3, 1750.

[96] ——. Application de la Machine Hydraulique de M. Segner à Toutes Sortes d'Ouvrages et de ses Avantages sur les Autres Machines Hydrauliques dont on se sert Ordinairement. *Mémoires de l'Académie des Sciences de Berlin*, 7:271–304, 1753. Presented to the Berlin Academy on September 2, 1751.

[97] ——. Théorie plus Complette des Machines qui sont Mises en Mouvement par la Reaction de l'Eau. *Mémoires de l'Académie des Sciences de Berlin*, 10:227–95, 1756. Presented to the Berlin Academy on September 13, 1753.

[98] Oliver Evans and Thomas Ellicott. *The Young Mill-wright & Miller's Guide. In five parts—embellished with twenty five plates*. The author, Philadelphia, 1795. [Knox Q4M .Ev1 and Wa. 174 @ 87191].

[99] Michael Faraday. *Chemical Manipulation; Being Instructions to Students in Chemistry, on the Methods Of Performing Experiments of Demonstration or of Research, with Accuracy and Success*. W. Phillips, London, 1827. [HK .F22 @ 216319].

[100] John Woodford Farlow. *The History of the Boston Medical Library.* Private printing by the Plimpton Press, Norwood, MA, 1918. [:XL5.B655.f @ 302310].

[101] Frank Felsenstein and James J. Connolly. *What Middletown Read: Print Culture in an American Small City.* University of Massachusetts Press, Amherst, MA, 1st ed., 2012. [Z1003.3.I6 F45 2015 @ 515361].

[102] Julie M. Fenster. *Ether Day: The Strange Tale of America's Greatest Medical Discovery and the Haunted Men Who Made It.* HarperCollins, New York, NY, 1st ed., 2001. [RD80.3 .F46 2001 @ 386886].

[103] Eugene S. Ferguson. *Engineering and the Mind's Eye.* MIT Press, Cambridge, MA, 1992.

[104] Jack Fergusson. The Periodic Table: A Historical Survey. *Chemistry Education in New Zealand*, pp. 1–12, May 2010.

[105] Massachusetts Society for Promoting Agriculture. Some Remarks on the Descructive Power of the Rose Bug. *Massachusetts Agricultural Repository and Journal*, 9(1):143–147, January 1826.

[106] Francis Fowler. Memoir of Dr. Thomas P. Jones. *The Franklin Journal*, 130:1–7, July 1890.

[107] James Bicheno Francis. *Lowell Hydraulic Experiments. Being a Selection from Experiments on Hydraulic Motors, on the Flow of Water over Weirs, in Open Canals of Uniform Rectangular Section, and Through Submerged Orifices and Diverging Tubes. Made at Lowell, Massachusetts.* D. Van Nostrand, New York, NY, 1868. [PJ/F84/2 @ 234305].

[108] Douglas A. Galbi. Book Circulation Per U.S. Public Library User Since 1856. *Public Library Quarterly*, 27(4):351–71, 2008.

[109] Abraham Gesner. *Remarks on the Geology and Mineralogy of Nova Scotia, Halifax, N.S.* Printed by Gossip and Coade, Halifax, Nova Scotia, 1836. [QE190 .G45 @ 10986].

[110] George Gibbs. Observations on the Dry Rot. *The American Journal of Science and the Arts (Silliman's)*, 2(6):114–118, November 1820.

[111] Owen Gingerich. *The Book Nobody Read: Chasing the Revolutions of Nicolaus Copernicus.* Walker & Company, New York, NY, 1st ed., 2004. [QB41 .G38 2004 @ 406210]. See also B.A.15 .20.124 for an audiocassette and DVD of a discussion held at the Boston Athenæum, March 25, 2004.

[112] James Whitbread Lee Glaisher. On the Progress to Accuracy of Logarithmic Tables. *Monthly Notes of the Royal Astronomical Society*, 33(5):330–45, March 1873.

[113] Samuel Goodrich. *Peter Parley's Tales about the Sun, Moon, and Stars: With Numerous Engravings*. Gray & Brown and Carter & Hendee, Boston, 1831.

[114] ———. *Peter Parley's Dictionary of Astronomy*. Otis, Broaders and Co., Boston, 1st ed., 1836. [:VEJ .G62 .d @ 292406]. Half-title: Parley's Cyclopedia. Dictionary of Astronomy.

[115] ———. *Peter Parley's Illustrations of Astronomy*. B.B. Mussey, Boston, 1840.

[116] ———. *Recollections of a Lifetime: Or Men and Things I Have Seen: in a Series Of Familiar Letters to a Friend, Historical, Biographical, Anecdotical, and Descriptive*. Miller, Orton & Co, New York, NY, 1st ed., 1857. [65 .G623 @ 135785].

[117] Maurizio Gotti. *Robert Boyle and the Langauge of Science*. Guerini Scientifica, Milan, Italy, 1996.

[118] ———. The Experimental Essay in Early Modern English. *European Journal of English Studies*, 5(2):221–39, 2001.

[119] Augustus A. Gould, Thaddeus William Harris, George B. Emerson, David Humphreys Storer, William Bourn Oliver Peabody, and Ebenezer Emmons. *Reports of the Commissioners on the Zoological Survey of the State*. Dutton and Wentworth, Boston, 1st ed., 1838. [LW64 .M38 @ 222385].

[120] Thomas Gray. *Observations on a General Iron Rail-Way: (With Plates and Map Illustrative of the Plan): Showing its Great Superiority, by the General Introduction of Mechanic Power, over all the Present Methods of Conveyance by Turnpike Roads and Canals, and Claiming the Particular Attention of Merchants, Manufacturers, Farmers, and, Indeed, Every Class of Society*. Baldwin, Cradock, and Joy, London, 1823. [PV .G79 @ 233810].

[121] Stephen Greenblatt. *The Swerve: How the World Became Modern*. W.W. Norton, New York, NY, 1st ed., 2011. [PA6484 .G69 2011 @ 480895].

[122] Scott B. Guthery. *A Motif of Mathematics: History and Application of the Mediant and the Farey Sequence*. Docent Press, Boston, 2010.

[123] ———. Raymond Clare Archibald and the Provenance of Mathematical Tables. In *Proceedings of the 38th Annual Meeting of the Canadian Society for History and Philosophy of Mathematics*, vol. 25, pp. 88–111. Canadian Society for History and Philosophy of Mathematics, 2012.

[124] Nathan Hale. *Inquiry into the Best Mode of Supplying the City of Boston with Water for Domestic Purposes; in Reply to the Pamphlets of Mr. Wilkins and Mr. Shattuck; and also to Some of the Representations to the Committee of the Legislature, on the Hearing of the Petition of the City.* Printed by Eastburn's Press, Boston, 1st ed., 1845. [PK964B6 .H @ 232535].

[125] William Hamilton. *The History of Medicine, Surgery, and Anatomy, from the Creation of the World to the Commencement of the Nineteenth Century.* Henry Colburn and Richard Bentley, London, 1831.

[126] Thaddeus William Harris. Minutes Towards a History of Some American Species Of Melolonthæ Particularly Injurious to Vegetation. *Massachusetts Agricultural Repository and Journal*, 10(1):1–12, July 1827.

[127] ———. *A Report on the Insects of Massachusetts, Injurious to Vegetation. Published agreeably to an order of the Legislature, by the Commissioners on the Zoological and Botanical Survey of the State. Insects of Massachusetts.* Folsom, Wells, and Thurston, printers to the University, Cambridge, 1st ed., 1841. [KZX .H24 .r @ 225525].

[128] ———. *A Treatise on Some of the Insects of New England which are Injurious to Vegetation.* John Owen, Cambridge, 1st ed., 1842. [SB931 .H @ 225527].

[129] ———. *A Treatise on Some of the Insects of New England which are Injurious to Vegetation.* Printed by White & Potter, Boston, 2nd ed., 1852. [KZX .H24 .t2 @ 225562].

[130] Thaddeus William Harris and Charles Louis Flint. *A Treatise on Some of the Insects Injurious to Vegetation.* William White, Printer to the State, Boston, 3rd ed., 1862. [KZX .H24 .t3 @ 225586].

[131] ———. *A Treatise on Some of the Insects Injurious to Vegetation.* Crosby and Nichols; O.S. Felt, Boston, New York, new ed., 1862. [KZX .H24 .t4 @ 225579].

[132] Ida Hay. George Barrell Emerson and the Establishment of the Arnold Arboretum. *Arnoldia*, 54(3):12–21, 1994.

[133] Thomas Wentworth Higginson. *Memoir of Thaddeus William Harris.* A. A. Kingman, Boston, 1st ed., 1869. [5 .9B v.24 no.6 @ 325923].

[134] Hina Hirayama. *With Éclat: The Boston Athenæum and the Origin of the Museum of Fine Arts, Boston.* Boston Athenæum, Boston, 1st ed., 2013. [N521.A8 H57 2013 @ 494278].

[135] Edward Hitchcock. *Report on the Geology, Mineralogy, Botany, and Zoology of Massachusetts. Made and Published by order of the Government of that State: In Four Parts: Pt. I. Economical Geology. Pt. II. Topographical Geology. Pt. III. Scientific Geology. Pt. IV. Catalogues of Animals and Plants. With a Descriptive List of the Specimens of Rocks and Minerals Collected for the Government.* J. S. and C. Adams, Amherst, 2nd ed., 1835. [$IK64 .M38 .2 @ 372850].

[136] Charles Hutton. *Mathematical Tables; containing the Common, Hyperbolic, and Logistic Logarithms; also Sines, Tangents, Secants, & Versed-Sines both Natural and Logarithmic. Together with several other Tables Useful in Mathematical Calculations. To which is Prefixed a Large and Original History of the Discoveries and Writings Relating to those Subjects; with the Complete Description and Use of the Tables.* F. C. and Rivington, London, 5th ed., 1811. [H2 .6H97 @ 224291].

[137] Charles Hutton and William Rutherford. *A Course of Mathematics, Composed for the Use of the Royal Military Academy.* T. Tegg, London, new ed., 1841. [H2 .H97 .2 @ 224661].

[138] Charles Loring Jackson. *Josiah Parsons Cooke. Biographical Notice*, Chapter 1, pp. 1–13. John Wilson and Son, Cambridge, MA, 1895. See also *Biographical Memoirs (National Academy of Sciences)* 4: 175–183.

[139] Charles T. Jackson. Notice of the Death of Francis Alger of Boston. *Proceedings of the Boston Society of Natural History*, 10:2–6, 1866.

[140] William B. Jensen. Physical Chemistry before Ostwald: The Textbooks of Josiah Parsons Cooke. *Bulletin for the History of Chemistry*, 36(1st):10–21, 2011.

[141] Walter R. Johnson. *Notes on the Use of Anthracite in the Manufacture of Iron: With some Remarks on its Evaporating Power.* C.C. Little and J. Brown, Boston, 1841. See also *Anthracite in the Manufacture of Iron.* Boston, 1847. [QC .J63 @ 236760].

[142] Thomas P. Jones. Abstract of the Report of the Committee on Premiums and Exhibition, on the Subject of the Third Annual Exhibition, qnd of Premiums Awarded. *The Franklin Journal*, 2:264–268, 1826.

[143] ———. List of Premiums Offered by the Franklin Institute of the State of Pennsylvania, and to be Awarded at their Third Annual Exhibition in 1826. *The Franklin Journal*, 1:6–10, 1826.

[144] ———. American Patents Granted in October, 1828. With Remarks and Exemplifications. *The Franklin Journal*, 6:393–401, 1828.

[145]   ———. List of American Patents Which Were Issued in October, 1829. With Remarks and Exemplifications. by the Editor. *The Franklin Journal*, 9:22–38, 1830.

[146]   ———. Specification of a Patent for an Improvement in the Construction of the Water Wheel, and in its Application to the Driving of Machinery, by the Reaction of Water. Granted to Calvin Wing, Gardiner, Maine, October 22, 1830. *The Franklin Journal*, 11:85–91, 1831.

[147]   ———. List of American Patents Which Issued in February 1834. With Remarks and Exemplifications, by the Editor. *The Franklin Journal*, 18:172–192, 1834.

[148]   ———. New Patent Law. *The Franklin Journal*, 22:158–66, 1836.

[149]   ———. The Patent Office. *The Franklin Journal*, 24:326–28, 1837.

[150]   ———. List of American Patents Which Were Issued in July, 1838. With Remarks and Exemplifications. by the Editor. *The Franklin Journal*, 27:11–23, 1839.

[151]   Louis Charles Karpinski and Walter Francis Shenton. *Bibliography of Mathematical Works Printed in America through 1850*. University of Michigan Press, Ann Arbor, MI, 1940. [Z6651 .K18 @ 25165].

[152]   Kimberly Kennedy. Uriah A. Boyden Papers. Smithsonian National Museum of American History Finding Aid, Washington, DC, 2010.

[153]   Anne Kelly Knowles and Chester Harvey. *Mastering Iron: The Struggle to Modernize an American Industry, 1800–1868*. University of Chicago Press, Chicago, IL, 1st ed., 2013. [HD9515 .K56 2013 @ 517471].

[154]   Nicolas Louis de La Caille, Joseph-François Marie, and Joseph Jérôme Le Français de Lalande. *Tables de Logarithmes pour les Sinus & Tangentes de toutes les Minutes du Quart de Cercle, & pour tous les Nombres Naturels depuis 1 jusqu'à 20000. Avec une Exposition Abrégée de l'Usage de ces Tables*. Chez Desaint, Paris, nouvelle ed., 1768. [Knox H4L .6L11 @ 94697].

[155]   Joseph-Louis Lagrange and Louis Poinsot. *Traité de la Résolution des Équations Numériques de tous les Degrés, aved des Notes sur Plusieurs Points de la Théorie des Équatins Algébriques*. Bachelier, Paris, 3rd ed., 1826. [H4E +L13 @ 226631].

[156]   James F. Lambert and Henry J. Reinhard. *A History of Catasauqua in Lehigh County Pennsylvania*. Searle & Dressler, Allentown, PA, 1914.

[157] Inés Lareo and Ana Montoya Reyes. Scientific Writing: Following Robert Boyle's Principles in Experimental Essays—1704 and 1998. *Revista Alicantina de Estudios Ingleses*, 20:119–37, 2007.

[158] Leslie Larson. *The Ropewalk at the Charlestown Navy Yard: A History and Reuse Plan*. Boston Redevelopment Authority, Boston, 1987.

[159] Charles T. Laugher. *Thomas Bray's Grand Design: Libraries of the Church of England in America, 1695–1785*. American Library Association, Chicago, IL, 1st ed., 1973. [Z731 .L38 1973 @ 91228].

[160] Edwin Layton. Mirror-Image Twins: The Communities of Science and Technology in 19th-Century America. *Technology and Culture*, 12(4):562–580, 1971.

[161] Edwin T. Layton. Millwrights and Engineers, Science, Social Roles, and the Evolution of the Turbine in America. In Wolfgang Krohn, Edwin T. Layton, and Peter Weingart, editors, *The Dynamics of Science and Technology: Social Values, Technical Norms and Scientific Criteria in the Development of Knowledge*, pp. 61–87. D. Reidel, Boston, 1978.

[162] ———. Scientific Technology, 1845–1900: The Hydraulic Turbine and the Origins of American Industrial Research. *Technology and Culture*, 20(1st):64–89, 1979.

[163] Ebenezer Avery Lester and Thomas P. Jones. Description Ebenezer A. Lester's Pendulum Steam Engine. *The Franklin Journal*, 12:98–104, 1831.

[164] I.W.P. Lewis. Examination – Light-House Establishment. Report upon the Condition of the Light-Houses, Beacons, Buoys, and Navigation, upon the Coasts of Maine, New Hampshire, and Massachusetts. Report 183, United States Department of the Treasury, Washington, DC, 1843.

[165] Nina Fletcher Little. Early Buildings of the Asylum at Charlestown, 1795–1846; now McLean Hospital for the Mentally Ill, Belmont, Massachusetts. *Old-Time New England*, 59(2):19–52, October-December 1968.

[166] Joseph L. Lord and Henry C. Lord. *A Defence of Dr. Charles T. Jackson's Claims to the Discovery of Etherization. Containing Testimony Disproving the Claims Set Up in Favor of Mr. W.T.G. Morton, in The Report of the Trustees of the Massachusetts General Hospital, and in No. 201 of Littell's Living Age*. Office of Littell's Living age, Boston, 1st ed., 1848. [RD80.J3 L8 @ 410328].

[167] J. C. Loudon, George Don, and David Wooster. *An Encyclopædia of Plants; Comprising the Description, Specific Character, Culture, History, Application in*

the Arts, and every other Desirable Particular Respecting all the Plants Indigenous, Cultivated in, or Introduced to Britain, Combining all the Advantages of a Linnean and Jussieuean Species Plantarum, an Historia Plantarum, a Grammar of Botany, and a Dictionary of Botany and Vegetable Culture. The Whole in English; With the Synonymes of the Commoner Plants in the Different European and other Languages. Printed for Longman, Rees, Orme, Brown and Green, London, 1st ed., 1829. [J .5L92 @ 219422].

[168] Paul Lucier. *Scientists and Swindlers: Consulting on Coal and Oil in America, 1820–1890.* Johns Hopkins University Press, Baltimore, MD, 2008.

[169] Patrick M. Malone. *Waterpower in Lowell: Engineering and Industry in Nineteenth-Century America.* Johns Hopkins University Press, Baltimore, MD, 1st ed., 2009. [TC425.M4 M35 2009 @ 471375].

[170] Gwen L. Martin. *Gesner's Dream: The Trials and Triumphs of Early Mining in New Brunswick.* Canadian Institute of Mining, Metallurgy and Petroleum, Fredericton, New Brunswick, 2003.

[171] William Matthews, William Henry Cox, and George Edward Madeley. *Hydraulia: An Historical and Descriptive Account of the Water Works of London and the Contrivances for Supplying other Great Cities, in Different Ages and Countries.* Simpkin, Marshall, and Co., Stationers' Hall Court, William Henry Cox, 5 Great Queen Street, Lincoln's-Inn Fields, London, 1st ed., 1835. [TD264.L8 M4 1835 @ 232779].

[172] Charles H. P. Mayo. *Memoir of Pestalozzi.* s.n., London, 1828. [B936 no.8 @ 352785].

[173] Haynes McMullen. The Very Slow Decline of the American Social Library. *The Library Quarterly: Information, Community, Policy,* 55(2):207–25, 1985.

[174] Martha J. McNamara. Defining the Profession: Books, Libraries, and Architects. In Kenneth Hafertepe and James F. O'Gorman, editors, *American Architects and Their Books to 1848,* pp. 73–89. University of Massachusetts Press, Amherst, MA, 2001.

[175] Domenico Bertoloni Meil. *Thinking with Objects: The Transformation of Mechanics in the Seventeenth Century.* Johns Hopkins University Press, Baltimore, MD, 2006.

[176] Solomon Miles and Thomas Sherwin. *Mathematical Tables; Comprising Logarithms of Numbers, Logarithmic Sines, Tangents, and Secants, Natural Sines,*

Meridional Parts, Difference of Latitude and Departure, Astronomical Refractions, &c. J. Munroe, Boston, stereotype ed., 1836. [H2 .6M59 @ 224293].

[177] Chrisopher Monkhouse. Parris' Perusal. *Old-Time New England*, 58(210):51–59, 1967. Published by Society for the Preservation of New England Antiquities, Boston, MA.

[178] Walter Scott Monroe. *Development of Arithmetic as a School Subject.* Government Printing Office, Washington, DC, 1917. Bulletin, 1917, No. 10 of the Bureau of Eduation, Department of the Interior.

[179] Arthur Morin. Experiments on Water-Wheels, having a Vertical Axis, Called Turbines. *The Franklin Journal*, 36:234–46, 289–302, 370–84, 1843.

[180] Ellwood Morris. Remarks on Reaction Water Wheels used in the United States; and on the Turbine of M. Fourneyron, an Hydraulic Motor, Recently used with the Greatest Success on the Continent of Europe. *The Franklin Journal*, 34:217–27, 289–304, 1842.

[181] ———. Experiments on the Useful Effect of Turbines in the United States. *The Franklin Journal*, 36:377–79, 1843.

[182] ———. On the Friction Dynamometer, or Brake, of M. de Prony, a Cheap, Simple, and Effectual Instrument, for Measuring the Actual Power Developed by Machines. *The Franklin Journal*, 35:225–38, 1843.

[183] Henry Moseley. *Illustrations of Mechanics.* Printed for Longman, Orme, Brown, Green, & Longmans, London, 1839. [QC127 .M67 1839 @ 226300].

[184] Sir Issac Newton. Scala Graduum Caloris. Calorum Descriptiones & Figna. *Philosophical Transactions of the Royal Society of London*, 22:824–29, 1701.

[185] Henry Robinson Palmer. *Description of a Railway on a New Principle: With Observations on those Hitherto Constructed: And a Table, Shewing the Comparative Amount of Resistance on Several Now in Use: Also an Illustration of a Newly Observed Fact Relating to the Friction of Axles, and a Description of an Improved Dynamometer, for Ascertaining the Resistance of Floating Vessels and Carriages Moving on Roads and Railways.* Printed for J. Taylor, London, 1824. [TF694 .P18 @ 10056].

[186] Antoine Parent. Sur la Plus Grande Perfection Possible des Machines, Étant Donnée une Machine qui ait pour Puissance Motrice quelque Corps Fluide que ce coit, comme, par exemple, l'Eau, le Vent, la Flamnie, &c. *Mémoires Académie des Sciences*, pp. 323–38, 1722.

[187] Richard Green Parker. *The Boston School Compendium of Natural and Experimental Philosophy, Embracing the Elementary Principles of Mechanics, Hydrostatics, Hydraulics, Pneumatics, Acoustics, Pyronomics, Optics, Electricity, Galvanism, Magnetism, Electro-Magnetism, Magneto-Electricity, and Astronomy, with a Description of the Steam and Locomotive Engines.* Marsh, Capen & Lyon, Boston, 1st ed., 1837. [HA .P22 @ 220605].

[188] Alexander Parris. *Plans of Buildings and Machinery Erected in the Navy-Yard Boston from 1830 to 1840.* Bureau of Yards and Docks, Washington, DC, 1838.

[189] Christine Pawley. *Reading on the Middle Border: The Culture of Print in Late Nineteenth-century Osage, Iowa.* University of Massachusetts Press, Amherst, MA, 1st ed., 2001. [Z1003.3.I7 P39 2001 @ 402431].

[190] Charles S. Peirce. Letter to the Editor. *The Nation*, 54(1392):1–2, March 1892.

[191] James Mills Peirce. *Three and Four Place Tables of Logarithmic and Trigonometric.* Ginn Brothers, Boston, 1874.

[192] Carole L. Perrault. Researching 19th-Century American Patents: The Journal of the Franklin Institute. *Bulletin of the Association for Preservation of Technology*, 8(2):24–56, 1976.

[193] William Phillips, Robert Allan, and Francis Alger. *An Elementary Treatise on Mineralogy: Comprising an Introduction to the Science.* W. D. Ticknor & Co, Boston, 5th ed., 1844. [I5 .P54 .t @ 217144].

[194] R Prony. *Nòuvelle architecture hydraulique, contenant l'art d'élever l'eau au moyen de différentes machines, de construire dans ce fluide, de le diriger, et généralement de l'appliquer, de diverses manieres, aux besoins de la société. Par M. de Prony.* Firmin Didot, Paris, 1 ed., 1790. [$PJ +P94 @ 368583].

[195] ———. *Recherches physico-mathématiques sur la théorie des eaux courantes / par R. Prony.* Imprimerie impériale, Paris, 1 ed., 1804. [HI +P94 @ 219680].

[196] Carroll W. Pursell. *Early Stationary Steam Engines in America: A Study in the Migration of a Technology.* Smithsonian Institution Press, Washington, DC, 1st ed., 1969. [Q5 .P977 @ 235727].

[197] Josiah Quincy. *The History of the Boston Athenæum, with Biographical Notices of its Deceased Founders.* Metcalf and Company, Cambridge, MA, 1851. [Z733 .B742 @ 49741].

[198] Terry S. Reynolds. Scientific Influences on Technology: The Case of the Overshot Waterwheel, 1752–1754. *Technology and Culture*, 20(2):270–95, 1979.

[199] ———. *Stronger than a Hundred Men: A History of the Vertical Water Wheel*. Johns Hopkins University Press, Baltimore, MD, 1983.

[200] A. W. Richeson. Warren Colburn and His Influence on Arithmetic in the United States. *National Mathematics Magazine*, 10(3):73–79, 1935.

[201] John Richie, Lindsy B. Schell, and Clarence H. Carter. *One Hundred Years of the English High School of Boston*. Centenary Committee of the English High School Association, Boston, 1924. [DQ64B6.En.o @ 78683].

[202] Evald Rink. *Technical Americana: A Checklist of Technical Publications Printed before 1831*. Kraus International Publications, Millwood, NY, 1st ed., 1981. [Z7912 T45 .R56 @ 12430].

[203] John Robison. *A System of Mechanical Philosophy*, vol. 2. John Murray, London, 1822. [HA .R56 @ 220696].

[204] Alan Rogers. Murder in Massachusetts: The Criminal Discovery Rule from Snelling to Rule 14. *The American Journal of Legal History*, 40(4):438–54, 1996.

[205] William Rosen. *The Most Powerful Idea in the World: A Story of Steam, Industry, and Invention*. Random House, New York, NY, 1st ed., 2010. [TJ461 .R67 2010 @ 471386].

[206] W.T. Russell. William Prescott Dexter. *Proceedings of the American Academy of Arts and Sciences*, 27:363–67, 1891.

[207] Carl Seaburg and Stanley Paterson. *Merchant Prince of Boston, Colonel T. H. Perkins, 1764–1854*. Harvard University Press, Cambridge, MA, 1st ed., 1971. [CT275.P472 S41 @ 3847].

[208] Secretary of the Scientific Library. *Record of the Proceedings of the Massachusetts Scientific Library Association: Proprietors of the Third Social Library in the City of Boston, 1826 Jan. 6–1827 March 17*. mss, Boston, 1826. [Mss..L10 @ 456998]. Minutes of meetings, which were held in the room of the American Academy of Arts and Sciences at the Boston Athenaeum on Pearl Street. The Association collected science books, with a preference for books on "mechanics and their various applications, particularly in civil engineering in the construction of roads and canals and the application of steam, water and wind to machinery - and to those on commerce, political economy and statistics ...history and literature generally to be excluded." The Association dissolved upon merging with the Boston Athenaeum in March 1827.

[209] Steven Shapin. Pump and Circumstance: Robert Boyle's Literary Technology. *Social Studies of Science*, 14(4):481–520, 1984.

[210] William Smith Shaw. *Memoir of the Boston Athenaeum with the Act of Incorporation, and Organization of the Institution*. Printed at the Anthology Office, Court-Street, by Munroe & Francis, Boston, 1st ed., 1807. [Z733.B74 M45 @ 51572].

[211] Thomas Sherwin. On Teaching the Elements of Mathematics. In *The Introductory Discourse and the Lectures delivered before the American Institute of Instruction, in Boston, August, 1834*, pp. 139–66, Boston, 1835. American Institute of Instruction, Carter, Hendee, and Co.

[212] ———. *An Elementary Treatise on Algebra: For the Use of Students in High Schools and Colleges*. Benjamin B. Mussey, Boston, 1st ed., 1842. [H4 .Sh5 .e @ 218188].

[213] ———. *The Common School Algebra*. Fred'k A. Brown & Co., Boston, 1862. [H4 .Sh5 @ 227604].

[214] Thomas Sherwin, Thomas Hill, and J. Giles. *Report of the Committee of the Overseers of the University at Cambridge, for Examining the Students in Mathematics: With a Report from a Minority of the Same Committee*. Metcalf and Company, Cambridge, MA, 1848. "Privately printed by order and for the use of the overseers. Not published." Signed: Thomas Sherwin. Report of the minority group signed: Thomas HIll, J. Giles.

[215] Benjamin Silliman. On the Iron Works at Vergennes, Vermont. *American Mineralogical Journal*, 1(1):80–83, 1810.

[216] ———. Particulars Relative to the Lead-Mine near Northampton, (Massachusetts). *American Mineralogical Journal*, 1(1):63–70, 1810.

[217] Catharina Slautterback. *Designing the Boston Athenæum: 10 1/2 at 150*. Boston Athenæum, Boston, 1st ed., 1999. [Z733.B74 S57 1999 @ 374673].

[218] John Smeaton. *A Narrative of the Building and a Description of the Construction of the Edystone Lighthouse with Stone: To which is Subjoined an Appendix Giving Some Account of the Lighthouse on the Spurn Point Built upon a Sand*. G. Nicol, London, 2nd ed., 1793. [TC375 .S63 1793 @ 25272].

[219] David Eugene Smith. The Development of American Arithmetic. *Educational Review*, 52:109–118, September 1916.

[220] David Eugene Smith and Jekuthiel Ginsburg. *A History of Mathematics in America Before 1900*. Open Court, Chicago, IL, 1934.

[221] Norman Smith. *Man and Water: A History of Hydro-Technology*. Charles Scribner's Sons, New York, NY, 1975.

[222] Neil Snodgrass. A Method of Heating Rooms by Steam. *The Repertory of Arts, Manufactures, and Agriculture*, 12:37–47, 1808.

[223] Arthut T. Stafford and Edward Pierce Hamilton. The American Mixed-Flow Turbine and its Setting. *Transactions of the American Society of Civil Engineers*, 85:1237–1356, 1922. Paper No. 1503.

[224] Charles S. Storrow. *A Treatise on Water-Works for Conveying and Distributing Supplies of Water; with Tables and Examples*. Hilliard, Gray and Co., Boston, 1835. [PK .St7 @ 232375].

[225] Theodore Strong. Remarks on Mr Colburn's Theory of Parallel Lines. *Boston Journal of Philosophy*, 3:371–372, 1826.

[226] James M. Swank. *History of the Manufacture of Iron in all Ages, and Particularly in the United States from Colonial Times to 1891. Also a Short History of Early Coal Mining in the United States and a Full Account of the Influences wich Long Delayed the Development of All American Manufacturing Industries*. American Iron and Steel Association, Philadelphia, 18892.

[227] James Joseph Sylvester. Elementary Researches in the Analysis of Combinatorial Aggregation. *The London, Edinburgh, and Dublin Philosophical Magazine and Journal of Science*, 24:285–296, 1844.

[228] Michael Taylor. *Tables of Logarithms of All Numbers from 1 to 101000; and of the Sines and Tangents to Every Second of the Quadrant. With a Preface and Precepts for the Explanation and Use of the Same, by Nevil Maskelyne*. Printed by C. Buckton, London, 1792. [H4L /6T21 @ 226633].

[229] Tamara Plakins Thornton. *Cultivating Gentlemen: The Meaning of Country Life among the Boston Elite, 1785–1860*. Yale University Press, New Haven, CT, 1989. [F73.44 .T48 1989 @ 74124].

[230] ———. *Nathaniel Bowditch and the Power of Numbers: How a Nineteenth-Century Man of Business, Science, and the Sea Changed American Life*. University of North Carolina Press, Chapel Hill, NC, 1st ed., 2012. [CT275.B67 T46 2016 @ 520309].

[231] David Bates Tower. *Intellectual Algebra, or, Oral Exercises in Algebra: For Common Schools in which all the Operations are Limited to Such Small Numbers as not to Embarrass the Reasoning Powers, but, on the Inductive Plan to Lead the Pupil Understandingly Step by Step to Higher Mental Efforts*. B.B. Mussey, Boston, 1st ed., 1845. [H4 .T65 @ 218195].

[232] ———. *Gradual Lessons in Oral and Written Arithmetic. Part I. Thought Combined with Practice. Take the First Step Right.* Tappan, Whittemore and Mason, Boston, 1850.

[233] Daniel Treadwell. *Report Made to the Mayor and Aldermen of the City of Boston: On the Subject of Supplying the Inhabitants of that City with Water.* Printed by True & Greene, Boston, 1st ed., 1825. [PK964B6 .T7 PK964B6 .T7 .2 @ 232547].

[234] ———. Description of a Machine, Called a Gypsey, for Spinning Hemp and Flax. *Memoirs of the American Academy of Arts and Sciences*, pp. 348–69, 1833.

[235] Thomas Tredgold. *Principles of Warming and Ventilating Public Buildings, Dwelling Houses, Manufactories, Hospitals, Hot-Houses, Conservatories, &c.; and of Constructing Fire-Places, Boilers, Steam Apparatus, Grates, and Drying Rooms; With Illustrations Experimental, Scientific, and Practical. To Which are added, Remarks on the Nature of Heat and Light; and Various Tables Useful in the Application of Heat.* Printed for J. Taylor, London, 2nd ed., 1824. [TH7010 .T779 @ 10101].

[236] James Russell Trumbull and Seth Pomeroy. *History of Northampton, Massachusetts, from its Settlement in 1654.* Printed by Gazette Printing Co., Northampton, 1st ed., 1898. [964N8 .T @ 166252].

[237] Trustees. Agricultural Experiments. *Massachusetts Agricultural Repository and Journal*, 10(1):89–90, July 1827.

[238] ———. Historical Sketch of the Massachusetts Society for Promoting Agriculture from 1792–1857. *Transactions of the Massachusetts Society for Promoting Agriculture*, 1:5–149, 1858.

[239] Edward Tuckerman. *An Enumeration of North American Lichenes, with a Preliminary View of the Structure and General History of these Plants, and of the Friesian System; to which is Prefixed, an Essay on the Natural Systems of Oken, Fries, and Endlicher.* J. Owen, Cambridge, MA, 1845.

[240] Edward Turner. *Elements of Chemistry: Including the Recent Discoveries and Doctrines of the Science.* John Taylor, London, 2nd ed., 1828. [HK .T85 @ 216979].

[241] United States Senate. *Public Documents Printed by Order of the Senate of the United States, First Session of the Twenty-Eigth Congress, Begun and Held at the City of Washington, December 4, 1843, in the Sixty-eighth year of the Independence of the United States.* Printed by Gales and Seaton, Washington, DC, 1844.

[242] Peter H. von Bitter. Abraham Gesner (1797–1864), An Early Canadian Geologist—Charges of Plagiarism. *Geoscience Canada*, 4(2):97–100, 1977.

[243] ———. Charles Jackson, M.D. (1805–1880) and Francis Alger (1807–1863). *Geoscience Canada*, 5(2):79–82, 1977.

[244] Clarence Abiathar Waldo. President's Address. *Proceedings of the Indiana Academy of Science*, pp. 35–53, 1899.

[245] William Waring. Investigation of the Power of Dr. Barker's, Mill, as Improved by James Rumsey, with a Description of the Mill. *Transactions of the American Philosophical Society*, 3:185–93, 1793.

[246] Edward Warren. *Some Account of the Letheon: or, Who is the Discoverer?* Dutton and Wentworth, Boston, 1847.

[247] R.C. Waterston. *Memoir of George Barrell Emerson, LL.D. Presented at the Meeting of the Massachusetts Historical Society, May 10, 1883.* John Wilson and Son, Cambridge, MA, 1st ed., 1884. [65 .Em33 .w @ 134530].

[248] John W. Webster. Notice of the Mineralogy of Nova Scotia, and of Several New Locations of American Minerals. *The Boston Journal of Philosophy and the Arts*, 3, 1826.

[249] Sara E. Wermiel. An Architect and Engineer in the Early Nineteenth Century: Alexander Parris's Engineering Projects. *Newsletter of the New England Chapters of the Society for Industrial Archeology*, 26(1):17–23, 2005.

[250] William Whewell. *The Philosophy of the Inductive Sciences: Founded upon their History.* John W. Parker, London, new ed., 1847. [H11 .W57 .p @ 224245].

[251] Frank P. Whitman. The Beginnings of Laboratory Teaching in America. *Science*, 8(190):201–6, 1898.

[252] Nicholas Wood. *A Practical Treatise on Rail-Roads, and Interior Communication in General; with Original Experiments, and Tables of the Comparative Value of Canals and Rail-Roads, and the Power of the Present Locomotive Engines.* Knight and Lacey, London, 1825. [PV.W85 @ 233878].

[253] Apollus Woodward. *A Brief Description of the Property Belonging to the Lycoming Coal Company with Some General Remarks on the Subject of the Coal and Iron Business.* Printed by P. Potter, Poughkeepsie, NY, 1828.

[254] David Wootton. *The Invention of Science: A New History of the Scientific Revolution.* HarperCollins, New York, 2015.

[255] Jeffries Wyman and William Healey Dall. *Biographical Memoir Of Augustus Addison Gould, 1805–1866*. National Academy of Sciences, Washington, DC, 1905.

[256] Morrill Wyman. *Practical Treatise on Ventilation*. Metcalf and Co., Cambridge, MA, 1st ed., 1846. [PF .W98 @ 232296].

[257] ———. Daniel Treadwell, Inventor. *The Atlantic Monthly*, 32:470–82, October 1873.

[258] ———. Memoir of Daniel Treadwell. *Memoirs of the American Academy of Arts and Sciences*, 11(6):325–24, 1888.

[259] Morrill Wyman, Jr. *The Early History of the McLean Asylum for the Insane : A Criticism of the Report of the Massachusetts State Board of Health for 1877*. Riverside Press, Cambridge, MA, 1877.

[260] ———. *A Brief Record of the Lives and Writings of Dr. Rufus Wyman [1778–1842] and his Son Dr. Morrill Wyman [1812–1903]*. Fourth March, Cambridge, MA, 1913.

[261] W. Ross Yates. Discovery of the Process for Making Anthracite Iron. *The Pennsylvania Magazine of History and Biography*, 48:206–223, 1974.

[262] Ronald J. Zboray. Antebellum Reading and the Ironies of Technological Innovation. *American Quarterly*, 40(1st):65–82, 1988.

[263] ———. *A Fictive People: Antebellum Economic Development and the American Reading Public*. Oxford University Press, New York, NY, 1993. [Z1003.2 .Z26 1993 @ 485430].

[264] Ronald J. Zboray and Mary Saracino Zboray. Books, Reading, and the World of Goods in Antebellum New England. *American Quarterly*, 48(4):587–622, 1996.

[265] ———. "Have You Read…?": Real Readers and Their Responses in Antebellum Boston and its Region. *Nineteenth-Century Literature*, 52(2):139–70, 1997.

[266] ———. Reading and Everyday Life in Antebellum Boston: The Diary of Daniel F. and Mary D. Child. *Libraries & Culture*, 32(3):285–323, 1997.

[267] ———. Home Libraries and the Institutionalization of Everyday Practices Among Antebellum New Englanders. *American Studies*, 42(3):63–86, 2001.

Index

The Readers entry in the following index contains the list of readers in experimental philosophy at the Boston Athenæum mentioned in the text. Entries in this list also appear in their proper alphabetic position in the index. The Quotes entry in the index contains the list of individuals who are quoted in the text.